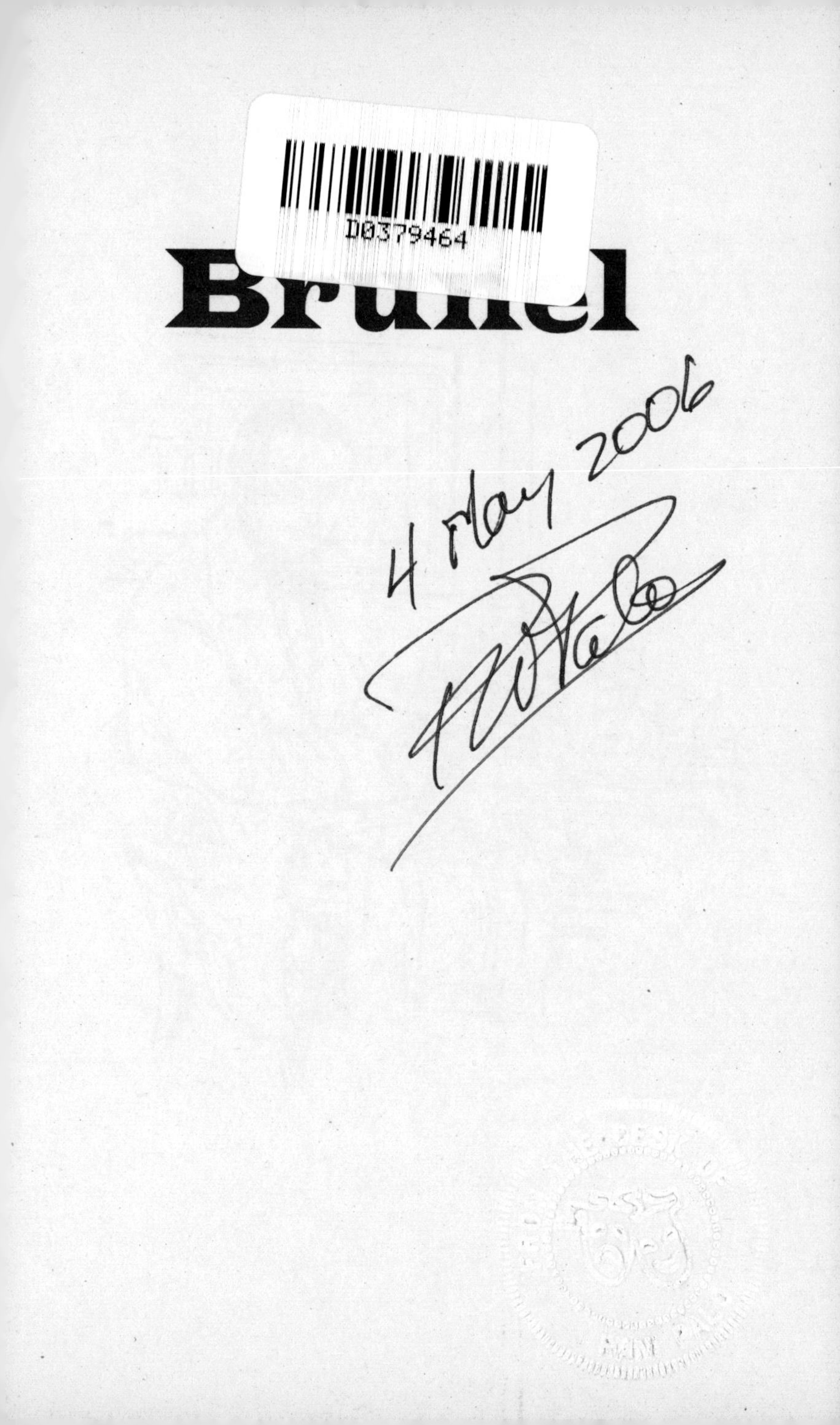

D0379464
Brunel
4 May 2006

AVON

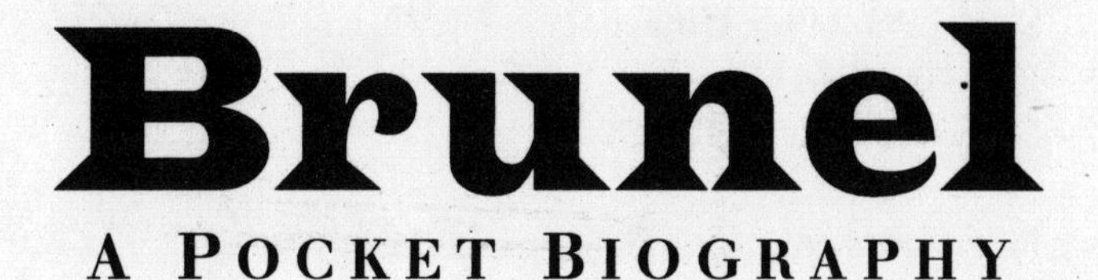

Brunel

A Pocket Biography

L.T.C. ROLT

Introduction by
SONIA ROLT

Illustrated by
PAUL SHARP

SUTTON PUBLISHING

Sutton Publishing Limited
Phoenix Mill · Thrupp · Stroud
Gloucestershire · GL5 2BU

First published 1965 by The Trinity Press, Worcester

This edition first published by Sutton Publishing, 2006

British Library Cataloguing in Publication Data
A catalogue for this book is available from the British Library.

ISBN 0-7509-4294-0

Printed and bound in England by
J.H. Haynes & Co. Ltd, Sparkford.

For
RICHARD
and
TIMOTHY

Contents

Introduction

FEW PEOPLE WILL OPEN THIS BOOK TODAY unaware of the heroic stature of Isambard Kingdom Brunel. When the book was written, nearly half a century ago, there were many who did not yet know his name; now we celebrate the great man's birth two hundred years ago, at the beginning of the century he was destined to change.

Brunel's brief span of fifty-three years was to be packed with action, and there to be were few before or since who have had a greater effect on the way we live ours lives today or on the innovations and designs which brought these changes.

This book was written by L.T.C., or Tom Rolt, who had already, eight years earlier, written a long, well-researched biography which remains in print to this day. The author had learnt for that book of the tremendous adventures that a life such as Brunel's provides for the reader. He was invited to write this shorter and simpler version for an American publisher. He dedicated it to his two sons, Richard and Timothy, who were twelve and ten at the time and already hands-on pragmatists, structuring dangerous tree houses and tunnels and railway lay-outs and tunnels in their Gloucestershire orchard.

Throughout *Brunel: A Pocket Biography*, Brunel's many-sided creativity and work are touched upon, and fall into natural chapters. In Portsmouth the Brunels' presence is linked with the British hero Nelson (Brunel's father Marc was a French émigré), whose victory and death the year before Isambard's birth had united the nation as one. Now, certain needs for the tackle of each of the navy's ships were to be provided for by Marc Brunel's blockmaking machinery. This was installed at the chief naval supply yard at Portsmouth and would produce hundreds of thousands of running blocks by a series of mechanical processes, whereas before each one had to be laboriously made by hand.

When of age, the young Brunel went to France, where the best technical education could be obtained. Later he came to join his father on the great enterprise to bore and build the first safe tunnel under the Thames. At the age of nineteen Isambard had to take charge of the work and he wrote electrifyingly of escapes from near death as the Thames threatened and broke in, and of the tenacity and loyalty of the work team. Money ran out, however, as so often on these ambitious exercises, and Isambard was soon handling his own projects.

The Great Western Railway followed the winning of a famous competition for the crossing of the deep Avon Gorge at Bristol. The graceful suspension bridge Brunel designed was not to be completed until after his death, its boldness modified. His vision and hope for his great railway eventually lost the battle of the gauges, but now the pace of his work and enterprise built up as he had become so widely known. Not all enterprises were successful, not all executed without criticism, not all without the suspicion that on occasion he could bite off more than he could chew; but all were imbued with a sense of wonder and a certain greatness in that he had achieved so much, often against great odds.

After so much – the design, the organisation, the bringing through, the achievement (in the case of the railway, frequently arguing before Parliament), the problems in new design, the steam ships built, the records broken or speeds and regularity maintained, the invention of the atmospheric railway and his prefabricated hospital for the Crimea – finally Brunel still had his last confrontation, his giant ship, the *Great Eastern*. He struggled with this as he became ill and finally died.

By what seems a meaningful irony, the great ship's work remained to be done after his death. She was to be the great finally successful layer of the first transatlantic telegraphic cable, linking the world together, under the direction of Isambard's greatest friend, the railway engineer Daniel Gooch. Gooch said of Isambard, 'Great things are not done under those who count the cost.'

Sonia Rolt
November 2005

I

Childhood and Youth

IN THE SPRING OF THE YEAR 1806 THE NAVAL Dockyard at Portsmouth was one of the wonders of the world, a place that no intelligent Englishman, no distinguished foreign visitor could afford to miss. Kings and nobles, statesmen and men of letters all flocked to Portsmouth along with the scientists and engineers to see the wonderful new machines for making ships' pulley-blocks. In a world in which practically every article of use was made by hand methods, aided only by a few very simple and crude machines, nothing like this new block machinery had ever been seen before. Each machine had been designed to carry out a particular operation and they worked so well and so quickly that the six men in charge of them could produce as many finished blocks in a day as sixty men had been able to do before.

"Machinery so perfect appears to act with the happy certainty of instinct and the foresight of reason combined." So wrote one admiring visitor, the novelist Maria Edgeworth, little knowing that she had glimpsed the future world of machines that we know today.

The machines were the handiwork of Henry Maudslay, the greatest mechanic of the day, but their designer was a Frenchman, a loyalist who had fled, first to America and then to England, to escape the French Revolution. His name was Marc Isambard Brunel and he was the younger son of a prosperous farmer of Hacqueville in Normandy.

Soon after he arrived in England in 1799, Marc Brunel had married Sophia Kingdom, an English girl whom he had earlier met in France. So that he could superintend the installation of his new machines, they went to live in a little house in Portsea, close to the dockyard, and by 1806 they had two small daughters, Sophia and Emma. Then, while the world admired the wonderful new machines, Sophia Brunel presented their engineer with his only son. He was born in the early hours of the morning of April 9, 1806, and his proud parents christened him Isambard Kingdom.

Isambard was still only a baby when, his father's work at Portsmouth completed, the family moved to No. 10 Lindsey Row, Chelsea. Here Isambard spent his boyhood, and what a happy and exciting time it must have been. His home was a part of a seventeenth-century mansion built for James I's doctor and later rebuilt by Charles II's Great Chamberlain, Lord Lindsey.

The old house stood on the waterfront of what was still a quiet village quite secluded from the hurly-burly of London. On one side its windows overlooked the traffic on the river and on the other a walled garden. Expeditions to London were generally made by boat from the steps below the house, and in summer there was bathing from these steps, for the Thames at Chelsea was not then the black, polluted stream that we know.

In the winter there were wonderful parties; in the spacious rooms of No. 10 candles glowed and fires blazed in the wide marble fireplaces as the children danced or acted charades while their elders played whist. The two Brunel girls were as prim and well-behaved as "young ladies" were expected to be in those days, but convention was less hard on boys, though even if this had not been so it seems unlikely that anything could have subdued their brother. In any children's game he soon became the ring-leader, while he brought tremendous energy and enthusiasm to everything he undertook, be it work or play. These two qualities, the gift of leadership and tireless energy, would stay with Isambard throughout his life.

After the success of his block-making machines, fortune turned against Marc Brunel for a time. He built a factory equipped with machinery of his own design to make army boots, but in the peace that followed the battle of Waterloo his scheme failed because the Government would not give him any orders. He also built a large steam-driven sawmill at Battersea, but on the night of August 30, 1814, the mill was completely destroyed by fire. Although these two disasters practically ruined him, he managed to give his only son a much better education than most boys could expect at that time. Isambard had mastered Euclid by the time he was six years old and also drew extremely well, so his father decided that such talent was worth every sacrifice he could make.

Isambard's drawings pleased his father particularly. The boy had already said that he wanted to be an engineer like his father, and Marc Brunel always insisted that an ability to sketch quickly and accurately was as important to an engineer as a knowledge of the alphabet. This, he argued, not only helped the engineer to express his ideas on paper, but it trained his powers of observation. In his own youth, when he was training to enter the French Navy at Rouen, Marc Brunel had spent his spare time drawing buildings in that city, and now Isambard followed his example in Hove when he was sent to a preparatory boarding school there. He proved the truth of his father's argument and astonished his school-mates by correctly predicting the partial collapse of a new building that was going up opposite the school.

It was thought at that time that mathematics and science were better taught in French schools than anywhere else in the world, so when Isambard was fourteen his father sent him first to the College of Caen in Normandy and later to the Lycée Henri-Quatre in Paris. After this, the boy served two periods of apprenticeship, the first in Paris under Louis Breguet and the second in London at the engineering works of his father's old friend, Henry Maudslay. Louis Breguet was unquestionably the greatest maker of watches and scientific instruments of his day, while in mechanical engineering the reputation of the firm of Maudslay, Sons & Field was unsurpassed. Their high standards of workmanship had a great influence on young Isambard, and for the rest of his life only the very best was good enough for him. This often led him into difficulties. He would be criticized as extravagant because he insisted always on the finest materials and workmanship. He would saddle himself with an enormous load of work and responsibility because he could not bring himself to trust assistant engineers to maintain the high standard he set.

In 1824, Marc Brunel started working on the most ambitious and best known of all his undertakings – the driving of a tunnel under the Thames from Rotherhithe to Wapping. It was to be the first of its kind in the world, for the difficulties and dangers of tunnelling under water are very great. By this time young Isambard, now aged eighteen, had just completed his education and training. He became, in effect, his father's junior partner and the tunnel project gave him an opportunity to show what he could do.

2

The Thames Tunnel Adventure

THE GREAT CORNISH ENGINEER RICHARD TREVITHICK had earlier tried to drive a tunnel under the Thames. His Cornish miners dug a small pilot tunnel, or "driftway" as it was called, and had almost reached the opposite shore when the river broke in and flooded the tunnel so quickly that Trevithick and his men were lucky to escape with their lives. This was in January 1808. Trevithick wanted to try again, but other engineers said the job was impossible; the people who had put up the money to build the tunnel believed them and so the work was abandoned.

Trevithick had followed the usual mining practice of using timber to support the sides and roof of his small tunnel, and Marc Brunel realized that for such underwater work something much stronger and better was needed to protect the miners and prevent collapse.

It is said that it was the heavily armoured head of the "ship-worm" which bores through ships' timbers that gave him the idea of the massive iron tunnelling shield that he designed and patented in 1818. As the picture of one section of this shield shows, the miners worked in a series of compartments inside it, and even the face they were excavating was protected by a series of heavy wooden boards held in place by jack screws. The miner removed one board at a time, dug the clay out behind it, then replaced the board and extended the jack screws to hold it in place. Meanwhile, close behind the shield, bricklayers built the tunnel lining. When the miners had excavated behind every board, all the jack screws were retracted and the whole shield moved forward by big jacks which pressed against the last ring of brickwork.

Marc Brunel's friends were so impressed by this idea that the Thames tunnel scheme was revived, and in 1824 the Thames Tunnel Company was formed with the elder Brunel as Chief Engineer. The Brunel family then moved from Lindsey Row to a house in Bridge Street, Blackfriars. They were very sorry to leave Chelsea village for dingy, smoky Blackfriars, but they realized that Marc Brunel had to live near the tunnel works, and their new home was soon made very comfortable.

At the beginning of March 1825, work began near the church of St Mary's, Rotherhithe. This consisted of building a huge cylinder of brickwork upon an iron ring or "curb". Young Isambard watched his father lay the first brick and then laid the second himself. A perilous engineering adventure had begun. When the brickwork was complete, workmen began to dig away the ground inside and beneath the iron ring, and as they did so the brick cylinder, 50 feet in diameter and 42 feet high, very slowly sank until its top reached ground level. Into the shaft so formed the tunnelling shield was lowered; a hole was cut in the brickwork and through this the shield began its slow and laborious advance under the river in November 1825.

It had been forecast that there would be firm clay at tunnel depth, but the shield soon struck loose gravel and quantities of water began to pour into the workings.

The shield alone made it possible to continue, but the work became extremely difficult and dangerous. To make things worse, the air in the tunnel became so foul that it caused serious fevers and blindness. At the outset, young Isambard was appointed to assist William Armstrong, the engineer-in-charge, but Armstrong was one of the first to fall seriously ill and he resigned his post. Isambard succeeded him and in this way, although he was still only nineteen years old, he took full responsibility, under his father, for an engineering work that was exciting the wonder of Europe.

Isambard threw himself into the work so whole-heartedly that sometimes he stayed in the tunnel for thirty-six hours at a time, taking only a short nap on the bricklayers' stage behind the shield. For fear he might work himself to death, he was given three assistants. One of these died of fever and another was struck partially blind, but apart from one short spell of illness Isambard carried on undaunted as the great shield crept slowly under the river. The further it went the worse the conditions grew until it became clear that there was only a treacherous layer of gravel between the top of the shield and the bed of the river. Although the bricklayers followed up closely, there was always a terrible risk that the river would break in whenever a section of the shield was moved forward, for this left a small part of the roof between shield and brickwork temporarily unsupported.

This was what happened on the night of May 18, 1827. There was a sudden shout from one of the miners in the shield, and the next instant all other sounds were drowned by the roar of water as the river burst in. A great wave rushed through the tunnel carrying all before it, but fortunately all the men who were working in the shield, and Isambard, who was standing in the tunnel, managed to escape. They scrambled up the shaft to safety a few seconds before the wave swept away the lower part of the stairs.

No sooner had he reached the surface than Isambard heard a faint cry for help coming from the darkness below. Someone realized that it was an old man named Tillett who looked after the pumps at the shaft bottom. Without a moment's hesitation, Isambard seized a rope, slid down an iron tie rod and managed to tie the rope round Tillett's waist as he struggled in the water. As soon as rescuer and rescued had been hauled to the surface, Isambard counted his men and found to his relief that no one was missing.

After such a narrow escape and so thrilling a rescue, most men would feel they had earned a long rest. Not so Isambard. Only twenty-four hours later he was at the bottom of the Thames in a diving-bell inspecting the damage from above. Sure enough, there was a large hole in the river bed. The diving-bell was manœuvred over the hole in such a way that Isambard was able to stand with one foot on the end of the tunnel brickwork and the other on the tail of the shield, a feat he afterwards recorded in a graphic little drawing.

This exploit nearly ended in disaster, for when Isambard's companion, Pinkney, tried to do the same thing he lost his hold upon the diving-bell and almost disappeared through the hole to certain death in the drowned tunnel. With great presence of mind, Isambard extended his leg as far as he could and Pinkney just managed to seize hold of his foot. By a tremendous effort, Isambard was able to drag him back into the diving-bell. Then he pulled the communicating cord as a signal to ascend and both were drawn to the surface; but having safely landed the dripping Pinkney, he was soon at the bottom of the river again.

The hole was stopped by laying iron rods across it and

then plugging it with bags of clay. The work of pumping out the tunnel could then begin and Isambard was the first to reach the shield, first by boat and then by crawling over the mound of earth that the river had washed in. But let Isambard's diary tell the story; his words show how excited he was by the whole experience and they tell us a lot about his character.

"What a dream it now appears to me!" he wrote. "Going down in the diving bell, finding and examining the hole! Standing on the corner of No. 12! The novelty of the thing, the excitement of the occasional risk attending our submarine excursions, the crowd of boats to witness our works all amused – the anxious watching of the shaft – seeing it full of water, rising and falling with the tide with the most provoking regularity – at last, by dint of clay bags, clay and gravel, a perceptible difference. We then began pumping, at last reaching the crown of the arch – what sensations! . .

"I must make some little indian ink sketches of our boat excursions to the frames: the low, dark, gloomy, cold arch; the heap of earth almost up to the crown, hiding the frames and rendering it quite uncertain what state they were in and what might happen; the hollow rushing of water; the total darkness of all around rendered distinct by the glimmering light of a candle or two, carried by ourselves; crawling along the bank of earth, a dark recess at the end – quite dark – water rushing from it in such quantities as to render it uncertain whether the ground was secure; at last reaching the frames. . . ."

By "frames" Isambard means the shield. Each section of the shield was numbered so that "No. 12" refers to that part where the water broke in.

This inspection of the tunnel was a most perilous undertaking. There was a great risk that the river might break in again and had this happened there would have been no hope of escape because the tunnel was already so full of water that there was only just room for the boat to move along under the tunnel roof. But for Isambard risks simply spelt excitement and adventure; now and throughout his life he loved to live dangerously.

It was six months before order was completely restored in the tunnel and work could begin again. In November 1827, Isambard celebrated this triumph over disaster by holding a banquet in the tunnel for fifty guests. The tunnel walls were hung with crimson drapery, a band played and gas lamps blazed down on the long white table. But Isambard did not forget the men who had helped him; in the adjoining arch one hundred and twenty miners sat down to the feast.

This was a splendid occasion, but the victory was very brief. As the shield began to creep forward again conditions were as bad as ever and disaster threatened daily. On January 14, 1828, it struck again. This time Isambard was standing in the shield with the miners when the river burst in with even more terrifying suddenness and violence. It swept them all off the staging which collapsed, drowning six unfortunate men who were pinned beneath it. Isambard had a miraculous escape. Though trapped by a falling timber which badly damaged his knee and also caused internal injuries, he managed to free himself. Then, believe it or not, he paused to admire the spectacle of the water rushing in until a great wave swept him out of the tunnel and up the shaft where one of his assistants managed to grab him.

"I shan't forget that day in a hurry," he wrote afterwards, "very near finished my journey then; when the danger is over, it is rather amusing than otherwise – while it existed I can't say it was at all uncomfortable. . . . While exertions could still be made and hope remained of stopping the ground it was an excitement which has always been a luxury to me . . . When knocked down I certainly gave myself up, but I took it very much as a matter of course, which I had expected the moment we quitted the frames, for I never expected we should get out. The instant I disengaged myself and got breath again – all dark – I bolted into the other arch . . . I stood still nearly a minute. I was anxious for poor Ball and Collins, who I felt too sure had never risen from the fall we had all had and were, as I thought, *crushed* under the great stage. I kept calling them by name to encourage them, and make them also (if still able) come through the opening.

"While standing there the effect was – *grand* – the roar of the rushing water in a confined passage and by its velocity rushing past the opening was grand, *very grand.* I cannot compare it to anything, cannon can be nothing to it. At last it came bursting through the opening. I was then obliged to be off – but up to that moment, as far as my sensations were concerned, and distinct from the idea of the loss of the six poor fellows whose death I could not then foresee, kept there.

"The sight and the whole affair was well worth the risk and I would willingly pay my share, £50 about, of the expenses of such a 'spectacle'. Reaching the shaft, I was much too bothered with my knee and some other thumps to remember much."

This time there was to be no rushing back next day to inspect the damage for Isambard's injuries were serious and he had spent more than fourteen weeks at Bridge House before he was well enough to write this extraordinary account of his escape in his neglected diary. Meanwhile his father had set about stopping this second breach by the same means as before, and Isambard looked forward to starting work in the tunnel again as soon as he was well enough. But it was not to be, for although the breach was filled and the tunnel cleared for the second time, no more money could be raised to pay for further work. The shield was bricked up where it stood to make the works secure and the unfinished tunnel was opened to the public as a peep-show.

"The tunnel is now, I think, *dead*," wrote Isambard bitterly. "This is the first time I have felt able to cry at least for these ten years . . . It will never be finished now in my father's lifetime I fear."

We can sympathize with his feelings. He had thrown himself heart and soul into the work and now it seemed that all the effort, the hardship and the dangers so fearlessly met had been in vain. But in one respect he was proved wrong. His father *did* live to see the tunnel completed and to earn a knighthood as a result. Work on the tunnel began again in 1835 and it was opened in March 1842. But in this second chapter in the history of the tunnel Isambard played no part. He had already achieved fame elsewhere.

3

Opportunity at Bristol

FOR A YOUNG AND AMBITIOUS ENGINEER IT WAS an exciting world and nowhere was it more exciting than in England. Fifty years before Isambard was born the word "engineer" was scarcely known except in its military sense, yet by the time he had grown to manhood a generation of great engineers – John Smeaton, William Jessop, Thomas Telford, John Rennie, James Watt, Richard Trevithick and his own father – had won for themselves a fame hitherto reserved for architects, and for their special skill the dignity of a profession. These men had given Britain a system of canals and roads; they had given the world steam power. By using new materials, cast iron and wrought iron, they had built bridges and aqueducts upon a scale never known before.

While the two Brunels were burrowing under the Thames, Telford had completed his two soaring suspension bridges that spanned the Conway and the Straits of Menai with their mighty wrought-iron chains. Nearer at hand Rennie's two sons were building new London Bridge to their father's design. Most exciting of all there came from the north rumours of new steam-powered railways, for in 1825 George

Stephenson and his young son Robert – only three years older than Isambard – had completed the Stockton & Darlington Railway.

Isambard's one aim in life was to join the ranks of these great ones, and he threw himself so unsparingly into the tunnel work because he saw it was a stepping stone to fame. In the little time he could spare from his work he would dream of his future in his Rotherhithe lodgings and sometimes he set down these dreams in his diary. He wanted to be "the first engineer and an example for future ones". He surveyed a world filled with great opportunities and exclaimed, "What a field – yet I may miss it."

The abandonment of the tunnel works came as a terrible blow to these hopes, but his determination to succeed was too strong to be broken by any set-back, and as soon as he had fully recovered from the effects of his accident he began to try by every means to get himself employed as an engineer. He did succeed in getting various jobs to do in different parts of England, but they were all of a very minor kind and no substitute for the tunnel as an opportunity. Yet Isambard was not too proud to undertake them, for he wisely realized that every job he did, however small, added something useful to his experience.

These jobs also took him about the country a great deal, which was also a useful experience, and on one such journey in the autumn of 1831 he travelled for the first time in a railway train. This was on the Stephensons' Liverpool & Manchester Railway which had then been working for little more than a year. While the primitive train swayed and jolted along, Isambard drew a series of wavering circles and lines on a sheet of paper. When he got home, he put this in his diary and wrote: "I record this specimen of the shaking on the Manchester Railway. The time is not far off when we shall be able to take our coffee and write while going noiselessly and smoothly at 45 miles per hour – let me try."

After this experience, which obviously fired him with enthusiasm, Isambard *did* try very hard. More than once he applied for the post of engineer to a newly promoted railway, but his hopes were dashed. His applications were refused; he was thought to be too young, too inexperienced. This made him very depressed for he had realized that these new railway schemes which the success of the Liverpool & Manchester Railway had brought about offered the best chance of winning fame. But oddly enough, while his direct efforts to become a railway engineer failed, he was soon to reach this goal by way of an opportunity that occurred as a direct result of his accident in the tunnel.

As soon as he was well enough to travel, his parents sent him to Clifton to convalesce. Close by his lodgings was the deep gorge of the river Avon, and as he grew stronger he was soon scrambling about on its steep, rocky slopes, gazing down at the shipping far below as it moved between Bristol docks and the Bristol Channel.

For many years men had been talking about building a bridge over the gorge, and now a local committee had been formed which had just launched a competition for the best bridge design. No less a person than Thomas

Telford had been asked to judge the entries. No sooner did Isambard hear about this than he determined to enter the competition. He produced three designs which you may see today in the Great Western Railway Museum at Swindon. Each shows a great suspension bridge, its chains anchored in the rock, bridging the gorge from lip to lip in one splendid span. But once again his hopes were dashed for Telford rejected all the competition entries as unsound.

Faced with this failure the bridge committee felt that the only thing they could do was to ask Telford to design a bridge himself. Telford had rejected Isambard's designs because he thought that their span – much longer than that of his own great bridge over the Menai Strait – was too great. He thought such a bridge would sway about dangerously whenever the wind blew strongly through the gorge. So, although Telford was himself in favour of a suspension bridge, in his own design he reduced the length of the span very much by showing two very tall piers rising from the bottom of the gorge on each bank of the river. The chains supporting the main span were to be carried over the tops of these piers.

Now whereas Isambard Brunel was young and practically unknown, Thomas Telford was the most famous living civil engineer in the country. But this did not stop Isambard from criticizing Telford's design in print. He called it timid and said that if such a bridge was built it would be no credit to British engineering skill. The bridge committee, which had at first praised Telford's design, now had second thoughts. They decided to hold another competition and this time Telford would be a competitor and the judges two members of the Royal Society. Isambard entered for this another design for a suspension bridge in which he reluctantly gave way to public opinion to some extent by showing a large brick abutment, to be built on the Somerset slope of the gorge. This enabled him to reduce the length of the span from the 916 feet of the boldest of his earlier designs to 630 feet, but this would still be the greatest span of any bridge yet built anywhere in the world. After some hesitation, the judges gave him the victory, the bridge committee appointed him their engineer and work on the bridge began in June 1831.

Isambard was delighted. This was his first real piece of good fortune since the closure of the tunnel, but ill luck still dogged him and he was soon disappointed again. For it soon became clear that the bridge committee had not got nearly enough money to pay for the work. They made efforts to raise more but without much success so that the work, after dragging along slowly and by fits and starts, finally came to a stop when only the big abutment on the Somerset side had been built. It was the sad story of the tunnel all over again, and whereas his father was able eventually to complete his tunnel, Isambard's beautiful bridge – "my first child, my darling", as he called it – was never finished in his lifetime. As a tribute to his memory, some of his fellow engineers raised the money and completed the Clifton bridge after his death, though the result differs in some respects from the original design. Yet the conception was his so that this splendid, soaring bridge stands today as a memorial to the boldness of a young engineer.

When work on the bridge stopped, Isambard felt more down-hearted than ever, for when he had won the competition he had really felt that success had come at last. Yet his work on the bridge had not been in vain after all. It had greatly impressed the merchants of Bristol, and they employed him to carry out some important improvement works in the city's docks.

It was while he was doing this that a local friend told him of the scheme to build a railway from Bristol to London. A committee had been formed to pursue the idea and were prepared to appoint as their engineer whoever would undertake to survey the cheapest route for the new line. Isambard needed little persuading to go and see this committee. To be the engineer of a great railway was his dearest ambition, yet it is typical of him that he risked losing this golden opportunity because he would not, to curry favour with the committee, lower the standard of perfection he had set himself. No, he would not undertake to survey the cheapest route for the railway; he would survey only one route – the best.

Such boldness and self-confidence from so young a man astonished the committee. They did not know what to make of him; their opinions were divided and the appointment would have to be put to the vote. A tense period of two days followed while the committee considered the merits of other applicants. Then, on March 6, 1833, a messenger arrived with a note. It said that Isambard Kingdom Brunel had been appointed engineer of the Bristol Railway, soon to be called the Great Western. Later, his friend told him that he had been elected by one vote. Close going indeed as Isambard himself remarked, but he had achieved his amibition at last and now it was up to him to prove himself worthy of his good fortune.

4

Fame at Last

BRUNEL – AS WE WILL CALL HIM NOW THAT HE was well and truly launched upon his own career – lost no time in setting to work. Within days of his appointment he was riding through the limestone hills from Bristol to Bath and then eastwards again through Box as far as Corsham, deciding on the most promising route for the new railway. He soon fixed upon Temple Meads, then open meadowland, as the best site for the Bristol terminus and the Avon valley as the best, though by no means easy, course to Bath. Then he hurried off to decide the best route into London, leaving an assistant surveyor at work in the west. The railway committee had called for the complete survey by the end of May at the latest, so there was no time to be lost.

Though he employed assistant surveyors, they needed constant guidance, and the next few weeks were very hectic for Brunel. Even in this age of the motor car it would be a remarkable feat to complete a survey for a line more than a hundred miles long in a mere six weeks, but Brunel's pace was set by the horse. He seldom managed to snatch more than an hour's sleep at a time during those weeks. Long days in the saddle or travelling by coach were followed by nights spent at country inns poring over his plans and calculations by candlelight.

Although Parliament insisted on seeing detailed plans before they would consider authorizing construction of a new railway, they could give surveyors no special authority to enter private property in order to prepare such plans. This gave landowners who were hostile to railways a golden opportunity. On the Liverpool & Manchester and again on the London & Birmingham the Stephensons and their assistants were more than once attacked by organized gangs armed with pitchforks and cudgels who drove them off their masters' property and smashed their surveying instruments. This was one difficulty which Brunel seems to have been spared, for his diary makes no mention of any violent encounters of this kind and he was able to present his survey to the Bristol Committee in May as he had promised.

In August the Bristol Committee, accompanied by Brunel, set out by coach for London for a joint meeting with the similar Committee which had been formed in London. At this historic meeting it was agreed to seek parliamentary powers to build the new line under the proud title of the Great Western Railway. That it was Brunel who suggested this name instead of "the London & Bristol" is very likely, for he already foresaw that the line might extend further west into Devon, Cornwall and Wales. A Board of Directors was formed and the secretary of the London Committee was appointed secretary of the new Company. His name was Charles Alexander Saunders and he became a lifelong friend and ally of Brunel.

Until now, Brunel's home had still been his parents' house in Bridge Street, Blackfriars, although he travelled about so much that he was seldom there. But it was essential for an up-and-coming railway engineer to have a London headquarters of his own, preferably near the Houses of Parliament. So Brunel found himself a small house and office at No. 53 Parliament Street and engaged a clerk named Bennett to look after his business for him while he was travelling about the country. Brunel inspired great loyalty in those who worked for him or with him, and Bennett stayed with him until his master's death.

The little office in Parliament Street at once became a hive of activity, for there was much to be done. The survey Brunel had made in the spring showed alternative routes for the line, one through the Kennet Valley from Reading by Newbury, Hungerford and then to the Vale of Pewsey, and the other keeping north of the Wiltshire downs along the Vale of the White Horse by Swindon and Wootton Bassett.

Although the Kennet Valley line was later used to form a more direct route between London, Taunton and the far west, it was decided that the northern line was the better way of reaching Bath and Bristol, and Brunel was ordered to make a more detailed survey of it for laying before Parliament. This meant another spell of high-pressure work in the field so that his plans would be finished by November. If they were not ready by then they would miss the next session of Parliament.

Once again Brunel finished his work on time, yet so far as he was concerned the battle for the Great Western Railway had not yet begun. When the Bill and plans for a new railway were laid before Parliament the promoters faced a long and expensive ordeal. After a purely formal "first reading" in the House of Commons, the Bill's second reading was moved, debated and voted upon. If it passed this second reading the Bill was then referred to a special Committee who could reject it if they were not satisfied.

It was here that the real battle had to be waged against those who opposed the railway. The engineer responsible for the plans and estimates of cost had to be prepared to defend them against the attacks of counsel employed by hostile landowners or by canal and stage-coach owners. For hours these men would cross-question the engineer, trying to find some fault in his plans and estimates or to make him contradict himself, and whether the Committee passed the Bill depended to a very great extent on the skill with which the engineer defended himself. Even if the Bill survived this Committee stage it still had to be passed by the House of Lords before it received the formal Royal Assent and became law. Only then could the promoters start building their railways.

Because they had not yet raised enough money to build the whole line, the promoters of the Great Western only applied to Parliament for power to build the London–Reading and Bath–Bristol sections. Their Bill passed its second reading in March 1834. After a wordy battle lasting fifty-seven days it then passed the Committee only to be rejected by the House of Lords by forty-seven votes to thirty.

They had come so near to success that neither Brunel nor his associates were discouraged by this. They presented a new Bill covering the whole line from London to Bristol in 1835, and after another long battle in Committee lasting forty days the Bill passed both Houses and became law on the last day of August in that year.

The Great Western Railway could now be built, but the victory had cost the Company no less than £88,710 in legal and Parliamentary expenses, an enormous sum when you remember that a pound was worth far more then than it is today. But the victory was above all a personal triumph for Brunel. On one occasion he was cross-examined by opposition counsel for eleven days at a stretch, but although these opponents were professional word-spinners, whereas he was an engineer still in his twenties and with a reputation still to make, he was more than a match for them.

One of those who heard him said afterwards: "His knowledge of the country surveyed by him was marvellously great . . . He was rapid in thought, clear in his language, and never said too much or lost his presence of mind. I do not remember ever having enjoyed so great an intellectual treat as that of listening to Brunel's examination."

When he was not attending Parliament, Brunel was rushing about the country, for even before the battle for the Great Western had been won he had been asked to engineer other railways which were, or would eventually become, part of the Great Western system: the Oxford Branch, the Cheltenham & Great Western Union, the Bristol & Exeter and the Merthyr & Cardiff (Taff Vale) Railways.

He now had plenty of money, for he was earning the equivalent of £10,000 a year or more by today's values. He no longer travelled by coach or on horseback but in his own private carriage drawn by four horses. Typically, he had designed this carriage himself

so that it would carry his plans and surveying instruments, food, drink and a plentiful stock of the cigars which he now smoked incessantly, and so that he could sleep in it comfortably if need be. A long black vehicle, its appearance was soon to become familiar to railway workers all over the west of England who nicknamed it the "Flying Hearse". To save precious time, as his railways extended into the west, Brunel would have this carriage loaded on to a flat truck and carried as far as possible by train.

He seldom found time to go to bed; a short nap in a chair or in his carriage was enough. Once, while he was on a visit to the Taff Vale line, he was found one morning asleep in a chair with the ash of a complete cigar lying undisturbed on his chest. Yet neither this overwork nor his sudden rise to fame and wealth changed Brunel's character in the slightest, which shows us just how strong that character was. His parents and his old friends saw in him the same gay and light-hearted boy who had been the ring-leader at the children's parties in the old home at Lindsey Row.

A friend named Burke who lived opposite him in Parliament Street and saw him almost daily during the long fight for the Great Western Railway, said of him: "He could enter into the most boyish pranks and fun, without in the least distracting his attention from the matter of business . . . I believe that a more joyous nature, combined with the highest intellectual faculties, was never created and I love to think of him as the ever gay and kind-hearted friend of my early years." It is surely extraordinary that Brunel should carry so much responsibility and live for many years under a pressure and strain that few could bear and yet give his friends the impression that he had not a care in the world.

A successful railway engineer in those days not only had to possess plenty of courage and determination but also infinite patience and skill in overcoming legal problems and in dealing with property owners. Such qualities are rarely combined in one man. In the Thames tunnel, Brunel had shown his courage, but even his close friends may have felt that he was too headstrong and impatient to succeed in the railway world. He soon proved otherwise. Said Burke: "I have never known a man who, possessing courage which to many would appear almost like rashness, was less disposed to trust to chance or to throw away any opportunity of attaining his object . . . In the character of a diplomatist . . . he was as wary and cautious as any man I ever knew."

Brunel often asked Burke to keep him company on his journeys to the West of England and Wales where the engineer frequently had to speak at meetings held in support of railway schemes, and his friend afterwards remarked upon "the enormous popularity which he everywhere enjoyed". It is obvious that so rare a mixture of youth and gaiety with courage and great ability charmed everyone he met.

For months Brunel was too busy to keep his diary up to date, but on Boxing Night, 1835, as he sat late and alone by the fireside in his little house in Parliament Street thinking over all the excitements of the past year, he took it down from the shelf and wrote: "What a blank in my journal! And during the most eventful part of my life. When I last wrote in this book I was just emerging from obscurity. I had been toiling most unprofitably at numerous things – unprofitably at least at the moment. The Railway certainly was brightening but still very uncertain – what a change. *The Railway* now is in progress. I am their Engineer to the finest work in England – a handsome salary – £2000 a year – on excellent terms with my Directors and all going smoothly, but what a fight we have had – and how near defeat – and what a ruinous defeat it would have been. It is like looking back upon a fearful pass – but we have succeeded. And it's not this alone but everything I have been engaged in has been successful.

"I am just leaving 53 Parliament St. where I may say I have made my fortune or rather the foundation of it and have taken Lord Devon's house, No. 18 Duke St. – a fine house – I have a fine travelling carriage – I go sometimes with my 4 horses – I have a cab & horse, I have a secretary – in fact I am now somebody. Everything has prospered, everything at this moment is sunshine. I don't like it – it can't last – bad weather must surely come. Let me see the storm in time to gather in my sails.

"Mrs. B. – I foresee one thing – this time 12 months I shall be a married man. How will that be? Will it make me happier?"

"Mrs B." was Mary Horsley who lived with her parents at No. 1 High Row (now No. 128 Church Street) Kensington and whose brother, John Horsley the artist, became a close friend to Brunel and painted his portrait. They were married on July 5, 1836. By this time the Great Western Railway was under construction and Brunel could only spare time for a short honeymoon in Wales. Then he brought his bride back to his new house, No. 18 Duke Street, overlooking St James's Park, which was to be both his home and his business office for the rest of his life. This was Duke

Street, *Westminster*, not Duke Street, St James's. Brunel's old home, and indeed the whole street, was demolished long ago to make space for Government offices.

5

The Great Western Railway

THE MEN WHO BUILT BRITAIN'S CANALS WERE architects as well as engineers, but as the new profession of civil engineer became established and publicly recognized so it became more clearly defined and separate from the older architectural profession. The effect of this was noticeable when the railways began to be built. The new railway companies engaged an Engineer-in-Chief and an Architect who not only designed railway buildings such as the stations but also architectural "treatments" for tunnel mouths or important bridges.

Alone among the great railway engineers, Brunel objected to this division of responsibility and insisted upon being architect as well as engineer. Whereas most of the other railway engineers were self-taught and incapable of such versatility, Brunel's education and his skill in drawing enabled him to do this. It meant a lot of extra work for him, but apart from his own satisfaction he claimed that he saved his employers a lot of money. He always carried a sketch book with him on his travels. This contained sheets of squared paper, beautifully bound, and had a piece of sand-paper pasted on the inside cover which he used to keep his pencil sharpened to a fine point. In this book he would sketch very rapidly and accurately the design for a new station building, a tunnel portal or a bridge, and because these sketches were made on squared paper it was not difficult for his draughtsmen to translate them into large-scaled working drawings. A collection of these Brunel sketch books is now preserved in the library of Bristol University. Brunel was determined that the line from London to Bristol should be, as he had said, "the finest work in England" and these sketch books show how much care and thought he devoted to every detail.

There were to be two inclines of 1 in 100 near Wootton Bassett and Box, falling towards Bath, but with these exceptions Brunel had surveyed a superbly straight and level course for the new railway. For 71 out of the total of 118 miles of line there would be no gradient steeper than 1 in 1000. On this Brunel planned to achieve speeds far higher than anything known before, for at a time when most people thought that a speed of 35 m.p.h. was tempting providence, he correctly foresaw a not far distant day when high speed would attract passengers: when, instead of being frightened by speed, travellers would expect it and grumble if they did not get it. Yet Brunel realized that he could not hope to "sell" the public speed unless it could be combined with complete safety. For this reason he rejected the rail gauge of 4 feet 8½ inches which the Stephensons had chosen for the Liverpool & Manchester and the London & Birmingham lines for no better reason than that it happened to be the gauge of the Killingworth Colliery railway where George Stephenson's first locomotive ran.

With difficulty he managed to persuade the Great Western Railway Company to let him lay the new line to a gauge of 7 feet. He argued that on this broad gauge the trains would run much more steadily and safely at high speed, that the carriages could be larger and more comfortable and that it would be easier to design more powerful locomotives because of the greater space between the wheels and frames.

Brunel also gave much thought to the way the line should be laid: how the rails should be shaped and supported. The Stephensons were using a chaired rail, mounting the chairs on heavy stone blocks such as were used on early horse-drawn tramways. Brunel decided to use a hollow rail with a flat base spiked down to, and continuously supported by, massive baulks of timber lying lengthways in the ballast and tied together at intervals by wooden crosspieces or "transoms" as they were called. A small slice cut from one of Brunel's rails (they became known as "bridge rails") would resemble the letter U turned upside down and they were rolled from wrought iron because the steel rail was a long way in the future.

Rails were very expensive and had to be bought by the ton. By giving his rail continuous support he not only hoped to get a smoother ride but to save money because he could use a lighter rail. In fact, Brunel's track did not give a very smooth ride because it was too solid. The railway engineers only learned by experience that for comfort the track must be able to "give" slightly under the weight of the wheels. But Brunel's "baulk road", as it was sometimes called, did prove very safe in the event of a derailment. Whereas a derailed vehicle quickly smashed itself to pieces by bumping over stone blocks or cross sleepers, on Brunel's track its wheels would run harmlessly along the timber baulks until the train could be pulled up. On one occasion when the van at the back of a Great Western train came off the line, it ran in this way for miles and then re-railed itself at a level crossing.

It is important that a new railway should begin earning money as soon as possible and so, like other lines, the Great Western was opened in sections as the work went forward. The first length to be completed ran from Paddington to Twyford and this was opened for traffic in July 1839. The most important work on this part of the line was the beautiful Wharncliffe viaduct across the Brent Valley at Hanwell.

Just west of the temporary terminus at Twyford, the railway had to be carried over the Thames to Maidenhead, and because the river authorities had insisted that the waterway must not be blocked by a lot of piers, Brunel had designed a bridge with only one pier in the river and two of the longest and flattest arched spans ever to be built in masonry. Brunel's jealous rivals – and there were plenty of them by this time – declared that such a bridge was bound to collapse under the first train that tried to cross it, but they were disappointed: the graceful Maidenhead bridge has stood fast from that day to this.

The other big work on this next section was the long cutting, 60 feet deep and nearly two miles long, through the chalk of Sonning Hill, near Reading. In those days all such work had to be done by men and horses, and the weather during the autumn and winter of 1839 was so bad that the contractors who had undertaken to drive the cutting, faced with a morass of mud, gave up in despair. Brunel then took charge himself, and by setting an army of over a thousand navvies and nearly a hundred horses to finish the cutting, the first train was able to steam through to Reading in March 1840.

After this the going was easier. Farringdon Road (now called Challow) became the temporary terminus in July; Hay Lane, near Wootton Bassett, came next in December, and Chippenham had been reached by the end of May 1841. Meanwhile work had been going on simultaneously from the Bristol end of the line, but here progress was much slower because four big bridges had to be built and no less than seven tunnels driven. However, the line was opened from Bristol to Bath in August 1840, and to work the traffic the engines and carriages were either made in Bristol or brought there by sea.

To unite London with Bristol it was now only necessary to complete the line from Chippenham to Bath, but this was the most difficult section of all to build. There were deep cuttings, embankments and viaducts, but, above all, there was what Brunel's critics called "that monstrous and extraordinary, most dangerous and impracticable tunnel at Box." This was to be the longest railway tunnel yet attempted, a great straight shaft on a gradient of 1 in 100 driven for nearly two miles through the limestone hills that encircle Bath.

When Brunel had first proposed such a tunnel, one expert had calculated that if the brakes of a train failed as it entered the tunnel on the falling gradient, it would be travelling so fast by the time it got to the other end that its passengers would not be able to breathe. Another declared that the vibration of the trains would make the tunnel collapse and that no passengers would ever dare to travel through it. But in spite of all this head-shaking, Brunel went ahead and work on the tunnel was begun in September 1836. Eight shafts, one of them 300 feet deep, were sunk from the top of the hills down to rail level. Using black gunpowder, the navvies began to blast their way forward from the shaft bottoms, the rock and soil being hauled in buckets up the shafts by horse "gins", while steam pumping engines kept the workings clear of water.

For two and a half years the navvies fought their way slowly forward through the rock, using a ton of gunpowder and a ton of candles every week. Brunel himself frequently went down into the workings, and on one occasion when two headings met and were found to be truly in line he was so delighted that he pulled a ring from his finger and gave it to the foreman of the navvy gang whose descendants treasure it to this day.

Never since the building of the pyramids had man attempted works upon such a scale as this, and yet one young man had made himself responsible for every detail. It is small wonder that there were moments when even he doubted if the railway would ever be finished, when troubles and difficulties seemed about to overwhelm him.

It was at such a moment of depression that he wrote a private letter to his best friend on the railway, Charles Saunders, and said:

If ever I go mad, I shall have the ghost of the

> opening of the railway walking before me, or rather standing in front of me, holding out his hand, and when it steps forward, a little swarm of devils in the shape of leaky pickle-tanks, uncut timber, half-finished station houses, sinking embankments, broken screws, absent guard plates, unfinished drawings and sketches, will, quietly and quite as a matter of course and as if I ought to have expected it, lift up my ghost and put him a little further off than before.

Yet Brunel was never really discouraged. As these words show, even when things seemed at their worst he was strong enough to regard his troubles in a half-humorous way. And at last, in June 1841, Brunel's illusive ghost became a reality, setting all doubts at rest; for in that month the great tunnel at Box was finished and the first train from Paddington ran through to Bristol. It was a great victory but a very expensive one for the line had cost £6,500,000 to build, more than double Brunel's original estimate.

By this time Brunel's railway empire was growing fast. The first train from London to Bristol was able to continue as far as Bridgwater over the metals of the new Bristol & Exeter Railway, and Exeter itself, 194 miles from Paddington, was reached on May 1, 1844. A month later the branch line to Oxford was opened, and a year after this the first train ran over the new line from Swindon to Gloucester through Stroud. Parliament sanctioned the South Wales Railway which would extend the broad gauge metals from Gloucester along the seaboard of Wales to Milford Haven. In the fields beside the junction of the two lines at Swindon a new town was rapidly growing around the first of the great railway workshops. Work on the South Devon Railway from Exeter to Plymouth had begun under Brunel's direction, and there were plans afoot to extend the lines still further to the west towards their ultimate terminus at Penzance.

In a few crowded years, Brunel had achieved the fame and fortune he had dreamed about in his lodgings at Rotherhithe. Yet there was a cloud on the horizon. So long as he aimed his broad gauge lines due westwards all was well, but as soon as he began to drive northwards from the London–Bristol line he began to encounter bitter opposition from rival narrow gauge companies. The first of these clashes came at Gloucester where his broad gauge met the narrow lines of the Birmingham & Gloucester Railway. After much argument the two companies agreed to share the same track between Gloucester and Cheltenham and this became the first stretch of "mixed gauge", using a third rail for the 7 foot gauge. It was at Gloucester that the inconvenience and the extra expense of using two gauges first became obvious. But Brunel would not give up. He still hoped to win the whole country over to his broad gauge and in this desperate battle he realized that his best weapon was speed.

6

The Battle of the Gauges

WHEN THE TIME CAME TO ORDER STEAM LOCOMOTIVES for the new railway from builders in the North of England, Brunel made his first bad blunder and one that is difficult to explain. He told the makers that the engines must not exceed 8 tons on four wheels or 10½ tons on six and that the speed of the pistons in the cylinders must not exceed 280 feet per minute when the engine was travelling at thirty miles an hour. These conditions completely defeated the argument that larger and more powerful engines could be used on the broad gauge.

The largest driving wheels in use on the narrow gauge at that time were 5 feet 6 inches diameter, but in order to keep the piston speed down the makers of the first Great Western engines had to make wheels of seven or eight feet diameter. One extraordinary locomotive named *Thunderer* had ten foot wheels and a boiler carried on a separate carriage. With such big wheels the engines could certainly travel fast – *Thunderer* did sixty miles an hour on a test run – but because they had to be made so light they had not the power to pull a useful load.

To look after the new engines a Locomotive Superintendent would be needed, and a young man named Daniel Gooch wrote to Brunel applying for the situation. The two men met for the first time in August 1837, and Brunel engaged Gooch there and then. He could not have made a better choice. Gooch spent twenty-seven years as Locomotive Superintendent of the Great Western and then, after a short interval, he returned as Chairman of the Company, a position he held for twenty-four years until his death in 1889.

Young though he was, by the time he came to the Great Western Gooch had already had considerable railway experience, including a spell at the works of Robert Stephenson & Co., of Newcastle, the pioneer builders of locomotives who had made the famous *Rocket*.

As soon as the new locomotives began to arrive poor Gooch found himself in terrible trouble. His experience told him that they could never be a success, but his employers, knowing little or nothing about these new and strange machines, could not appreciate this. So, while he laboured day and night to try and keep the engines running, he had to take the blame for their continual breakdowns. In this desperate situation he learned that a locomotive he had helped to design for an American railway had never been delivered but was still lying unwanted in Robert Stephenson's shops at Newcastle. He knew that the gauge could readily be altered from the 5 feet 6 inch of the American line to 7 feet so, through Brunel, he persuaded the Company to buy it.

This engine was the famous *North Star*, a full-sized replica of which is to be seen in the Great Western Museum at Swindon. Too big and heavy to be delivered by canal boat to West Drayton like the other locomotives, *North Star* was brought by barge up the Thames to Maidenhead where she lay under a tarpaulin until the rails reached Twyford. She was the first successful broad gauge locomotive.

Even the Great Western directors had begun to have serious doubts about Brunel's broad gauge, but at the end of December 1839, *North Star* set their doubts at rest by hauling their special train from Paddington to Maidenhead at a start to stop average of 38 miles an hour. No one was more delighted by this success than Brunel. The Newcastle firm could not deliver sister engines quickly enough for him, and over the next few months *North Star* was joined by *Morning Star*, *Evening Star* and *Dog Star*.

These engines saved the reputation of the broad gauge, but they were only a beginning. Gooch had become as great a believer in the broad gauge as Brunel himself and the two men determined to show the world what it could do. Gooch set to work to design a larger and more powerful version of the "Stars" with 7 feet single driving wheels, and the first of these new engines, *Firefly*, was delivered in time to draw the directors' special when the line was opened to Reading in 1840. On the return journey, *Firefly* covered the 30¾ miles from Twyford to Paddington in 37 minutes, an average of 50 miles an hour and a feat of sustained high-speed travel such as the world had never known. Many engines of this new "Firefly" Class were built, and firmly established the broad gauge reputation for speed.

Meanwhile, as a result of the first encounter at Gloucester, the "Battle of the Gauges" was hotting up. The narrow gauge companies tried by every means to prevent Brunel from extending his broad gauge into the Midlands. They won the first round when the narrow gauge Midland Railway snatched the Birmingham to Bristol line from under the noses of the Great Western, in spite of the fact that the part between Gloucester and Bristol had been laid out by Brunel as a broad gauge line. The Great Western struck back by getting powers to build a broad gauge line from Oxford to Birmingham. The gauge question was argued so hotly all over the country that in 1845 the Government appointed a Royal Commission to decide on it.

This Commission accepted Brunel's proposal that the rivals should have a speed contest over two similar stretches of line: Paddington to Didcot, 53 miles, and Darlington to York, 44 miles. Matched against two new narrow gauge locomotives, Gooch's five years old *Ixion* of the "Firefly" class achieved 60 miles an hour with an 80-ton load and with 60 tons averaged 50 miles an hour over the 53 miles, start to stop. The narrow gauge challengers could not match this performance. One ignominiously ran off the line and turned over, while the highest speed the other could attain was 53¾ miles an hour with 50 tons.

Faced with these results the Gauge Commissioners could not do other than acknowledge broad gauge superiority in speed, yet they recommended the adoption of the narrow as the standard gauge for Britain. Brunel and Gooch were bitterly disappointed, yet the Commissioners' verdict was based on the facts that by this date there were nearly 2,000 miles of narrow gauge line in use against only 274 miles of broad gauge, and that it is much easier to narrow the gauge of a railway than it is to widen it. These facts spelled the doom of the broad gauge, yet Brunel refused to admit defeat and

planned, with Gooch's help, an even more convincing display of broad gauge superiority in 1846.

In the spring disturbing rumours began to reach the narrow gauge "enemy" that a giant locomotive was building in the new railway workshops at Swindon. Sure enough, at the end of April there rolled majestically out of the shops Gooch's *Great Western.* Though she might appear small to us today, in 1846 she dwarfed all other locomotives. Her single driving wheels were 8 feet in diameter and her big boiler carried steam at 100 lb per square inch – an unusually high pressure in those days. On June 1, *Great Western* was given her first run on the Exeter express and covered the 194 miles between Paddington and Exeter in 208 minutes down and 211 minutes up, an average of over 55 miles an hour. Later in the same month she hauled a hundred-ton train over the 77½ miles from Paddington to Swindon at a start to stop average of 59 miles an hour. These feats far surpassed anything hitherto achieved on rails and more than confirmed the claims that Brunel had made for his broad gauge.

The only fault with the *Great Western* was that the weight on the single pair of leading wheels was too great, so on the second engine of the type to be built, the *Iron Duke*, a second pair of wheels was added so that the wheel arrangement became 4–2–2–. *Iron Duke* became the name of a class, for she was the first of a whole series of broad gauge flyers with splendid names like *Lord of the Isles*, *Tornado*, *Typhoon* and *Lightning*. These engines hauled the first express trains in the world: the famous *Flying Dutchman* and the *Zulu* (both named after race-horses) on the Paddington to Exeter run.

Had Brunel been first in the field with his broad gauge he might have succeeded, but although broad gauge speed won him much support in the West Country he could not hope to win over all the narrow gauge lines that had been built in the Midlands and the North. The line that really defeated the broad gauge was the Oxford, Worcester & Wolverhampton Railway. This had been surveyed by Brunel as a broad gauge line and its construction was partly financed by the Great Western; but because it was also linked to the Midland and the London & North Western Railways it was agreed that it should be of mixed gauge.

But although some broad gauge rails were laid they were never used for regular traffic because the Company managed to evade its undertaking to run broad gauge trains. Then it joined forces with two other narrow gauge lines, the Worcester & Hereford and the Hereford, Abergavenny & Newport to become the West Midland Railway. In 1863, the West Midland was taken over by the Great Western, and because it was essential to work through traffic over this line from London, narrow gauge rails had to be laid inside the broad gauge between Oxford and Paddington.

Brunel died four years before "the coal wagon gauge", as his supporters called it, reached Paddington, but had he lived he would have realized that this spelled defeat, although on his old main line to the west the broad gauge was not finally abolished until 1892.

Commercially, the broad gauge was an expensive failure – it came too late; but as a feat of engineering it was a triumphant success. It also had a great influence on the progress of railways generally because it forced the narrow gauge to make improvements. As one writer put it at the time: "England owes much to Brunel for spurring on Stephenson; for had it not been for the Great Western, we should never have got the great speed which we now have."

7

The Atmospheric Railway and the Saltash Bridge

BRUNEL'S RAILWAY WORK TOOK HIM NOT ONLY to the west of England and South Wales, but to Ireland and to Italy where he surveyed new lines from Florence to Pistoja and across the Apennines from Genoa to Alessandria. So much work and travelling left him little time for home life, but his three children – Isambard, Henry Marc (who became an engineer like his father) and Florence Mary – loved him dearly. For them his homecomings were red letter days; they would listen eagerly for his return, knowing it would not be long before they heard his familiar tread climbing the nursery stairs two steps at a time. Then he would organize wonderful games for them, just as he had done in his own boyhood at Lindsey Row, or delight them with conjuring tricks.

It was during one of these nursery playtimes that Brunel came nearer to death than at any other time in his adventurous life. In doing a conjuring trick he accidentally swallowed a half-sovereign instead of just pretending to do so, and the coin lodged in his windpipe or "went down the wrong way" as we say. An operation to remove the coin failed and for some days he was in great danger of choking to death. Finally, Brunel saved himself by his own ingenuity. He designed a large board, pivoted between two uprights, on which he could be strapped down and spun round, head over heels. After one terrible fit of coughing centrifugal force succeeded where the surgeons had failed, and as he was whirled round Brunel felt the coin leave its place and drop harmlessly into his mouth.

Such was Brunel's fame and so great the anxiety caused by his accident that when the poet Macaulay ran through the Athenaeum Club shouting excitedly "It's out! It's out!" everyone knew what he meant. The event also found its way into the *Ingoldsby Legends* in the lines:

All conjuring's bad! They may get in a scrape
Before they're aware, and, whatever its shape,
They may find it no easy affair to escape.
It's not everybody that comes off so well
From "leger de main" tricks as Mr Brunel.

In the same year that Brunel met with this accident a very curious little railway indeed began working in Ireland. This was the Dalkey branch of the Dublin & Kingstown Railway. Opened on August 19, 1843, it was less than two miles long, but it was worked on what was called the atmospheric system. This had been developed and patented by an inventor named Clegg, helped by the brothers Jacob and Joseph Samuda who were Thames shipbuilders.

Their idea was this. A pipe was laid between the rails and the leading vehicle of each train carried a piston which ran inside this pipe. A large stationary steam engine pumped the air out of the pipe ahead of the train so that it was driven along the line by atmospheric pressure acting on the back of the piston or, in simpler language, it was sucked along. The idea was simple enough, but the practical problem was how to connect the piston to the carriage and still keep the pipe air-tight. There had to be a slot along the top of the pipe through which the piston arm could pass and this slot was sealed by what was called a continuous valve consisting of a leather flap weighted with iron. The leading carriage was equipped with special fittings to open the valve just ahead of the piston arm and to seal it down again just behind it.

The Dalkey branch was the first full-scale trial of this system. The line climbed continuously from the main line junction at Kingstown to its terminus at Dalkey, and a single pumping engine at Dalkey sucked the trains up this gradient. They came down again by gravity. The most famous railway engineers of the day came to see the railway working and formed very different opinions about it. The Stephensons and Joseph Locke were against it, but Cubitt, Vignoles and Brunel thought that the atmospheric system was worth trying out on a bigger scale in England.

At this time the South Devon Railway Company had been formed to extend the West of England main line from Exeter to Plymouth with Brunel as engineer. He had originally surveyed a route for this new line all along the Devon coast via Torquay, with big bridges over the estuaries of the rivers Teign and Dart, but this had to be abandoned as too expensive.

Instead, Brunel planned to carry his line inland up the valley of the Teign to Newton Abbot, then through Totnes and along the southern edge of Dartmoor. This would be a far cheaper line to build, but the gradients

would be much steeper, and Brunel doubted whether steam locomotives would be able to work trains quickly and reliably over such a route. So the South Devon directors accepted his suggestion that the atmospheric system should be used instead.

Brunel had been impressed by the way the big pumping engine sucked the trains up to Dalkey, swiftly and with no smoke, noise or smell such as a labouring steam locomotive would produce. Also, he reckoned that he could save the South Devon Company money on track and bridges because, instead of a heavy locomotive, the trains only needed a piston to draw them – all the power and weight was in the pumping engines at the lineside.

So it was that the atmospheric pipe was laid down from Exeter to Newton Abbot and on this stretch – which included that dramatic piece of line on the very edge of the sea from Dawlish Warren to Teignmouth – eight pumping stations were built, one for every three miles of pipe.

Atmospheric trains began to run regularly between Exeter and Teignmouth in September 1847, and the service was extended to Newton Abbot in January. At first the trains ran well; on one occasion a speed of 64 miles an hour was maintained for four miles, but very soon the troubles began. The chief trouble was with the continuous valve. The leather flap began to perish, causing the valve to leak, with the result that the pumping engines had to work so hard to overcome the leakage that the system became terribly expensive in fuel.

By June 1848, the leather had completely rotted away and Brunel was faced with the difficult choice between abandoning the atmospheric system altogether or renewing the whole length of the valve at an estimated cost of £25,000, a very large sum in those days. With characteristic courage, Brunel did not hesitate. Although several members of the South Devon Board were in favour of continuing the experiment, they accepted Brunel's advice to scrap the atmospheric system because it could never be a practical success.

The whole of the long pipeline was torn up, including the section over the steep inclines from Newton Abbot to Totnes which had been laid but never brought into use, and the victorious steam locomotives moved in. Brunel's idea of generating tractive power at the lineside had been right, but nearly half a century would pass before the development of electric power made its application successful on railways.

This costly failure of "the atmospheric caper", as the Devonians called it, was a bitter disappointment for Brunel and a mistake that could have ruined the career of many a lesser engineer. Yet his reputation was so great that it withstood the blow. Not only did he remain engineer of the South Devon Railway but of the Cornwall and the West Cornwall Railways that between them carried his broad gauge metals through to Penzance. It was on the Cornwall Railway that he carried out his last and greatest feat of railway engineering.

In order to carry the two Cornish railways and their branches over the many deep valleys that crossed their lines of route, Brunel designed a splendid series of timber viaducts on stone piers. One of these, St Pinnock, near Liskeard, was 153 feet high. They served their purpose well until increasing train weights and the difficulty of obtaining durable timber for repair led to their replacement either by masonry arches or by embankments. But there was one barrier that could not be crossed in this economical way. This was the deep tidal estuary of the Tamar at Saltash which forms the boundary between Devon and Cornwall, and here

Brunel built one of the world's most famous bridges.

The greatest of bridge builders before the age of railways was Thomas Telford. He became a master in the use of cast iron, and some of his iron road bridges stand to this day. His suspension bridge over the Menai Straits was the greatest work of its kind in the world when first completed. But neither cast-iron girders nor wrought-iron suspension chains were suitable for bridges of large span that had to support the weight of a railway train.

Many early railway bridges were built of cast iron, but Brunel mistrusted it as brittle and treacherous, and the collapse under a train of Robert Stephenson's cast-iron girder bridge at Chester showed that he was right. Similarly, the miserable failure of a suspension bridge carrying the Stockton & Darlington Railway over the Tees at Stockton showed that that type of structure was not suitable either.

In their search for an answer to this problem both Robert Stephenson and Brunel began to experiment with wrought-iron girders, and in this they were helped by the experience of the shipbuilders who were then just beginning to build large hulls in this material. Where Stephenson was concerned the outcome of these experiments was the great Britannia bridge over the Menai Straits where the trains pass through a huge rectangular tube of wrought iron. Meanwhile Brunel continued his experiments, searching for some simpler and more economical form of structure than Stephenson's great tube.

The bridge carrying the railway to South Wales

over the river Wye at Chepstow was the first really important outcome of Brunel's ideas and experiments. The single main span carrying the two tracks over the river consisted of two round tubes, each 9 feet in diameter and 300 feet long, carried on masonry piers and supporting the wrought-iron girders of the bridge deck below by means of tie rods and chains. The bridge served its purpose well and has only recently

been reconstructed by the same firm of contractors who built the original.

The crossing of the Tamar was a much bigger problem, for this river at Saltash is 1,100 feet wide and 70 feet deep at high water. Because it was used by ships of the Royal Navy, the Admiralty insisted that the bridge must be at least 100 feet above water level to clear their masts, and that the channel must not be obstructed by many piers.

Altogether, Brunel made four designs for the Saltash

bridge. The one he finally decided upon had two main spans of 465 feet each and therefore required only one pier in the middle of the river. It was originally designed to carry two tracks, like Chepstow, but this was judged too expensive and the bridge was built for a single line only.

The design of the two identical spans was upon the same principle as Chepstow but improved in detail. The huge wrought-iron tube supporting the bridge deck was oval instead of round and arched instead of almost straight. By making the tube broader (16 feet 9 inches) than it was high (12 feet 3 inches), Brunel made its width equal to that of the bridge deck below, so that the supporting chains and tie rods hung vertically instead of being splayed outwards as they were at Chepstow.

Standing to this day, the great Saltash bridge is the finest surviving example of Brunel's engineering genius and daring, yet the most difficult part of the whole achievement is hidden forever from our sight. This was the building beneath the water of the lower part of the central pier. In constructing his tubular bridge at the Menai, Robert Stephenson was able to build his central pier on a rock in mid-channel which was exposed at low tide, but for Brunel there was no natural help of this kind. To find a base for his pier he had to go down through eighty feet of water and mud. No engineer had faced as big a problem as this before.

Brunel's plan was to sink an iron cylinder down through the water and mud until it found a rock base and then build a pier of stone within it. The idea was first tried out on a small scale by sinking a trial cylinder 7 feet in diameter. When it reached the rock the water inside it was pumped out, trial borings were made to find out the nature of the rock and to plot the shape of its surface. Then a huge cylinder 35 feet in diameter was built by the river bank, towed out to mid-channel and then slowly and with great difficulty upended and sunk until it rested on the rock. This "great cylinder", as it was called, was really a tube within a tube. The space between the inner and outer tubes was divided into a series of compartments which could be filled with compressed air. Men entered these compartments through an air-lock to dig away the mud and so allow the cylinder to sink, while the compressed air kept the water out.

The stone pier would be built inside the inner tube. Brunel had designed this part of the great cylinder in the form of a diving-bell such as he had used many years before after the water had burst into the Thames Tunnel. He hoped that this, with the help of the air pressure in the outer compartments, would keep the water out. He did not want to use air pressure here if he could avoid it since this would mean that the masons and all their materials would have to pass through air-lock doors, making the work very slow. Also, men working under air pressure are apt to suffer from a very painful complaint called "the bends" if they return too quickly to a normal atmosphere.

Unfortunately, however, a large fissure or fault was found in the rock on which the great cylinder stood, and through this water poured in such quantity into the inner cylinder that no amount of pumping would keep it out. So, after all, the whole of the cylinder had to be pressurized. In order to do this it had to be strengthened by lowering into it the original small trial cylinder. This was a very difficult, delicate and dangerous operation, but it was successfully carried out and the building of the pier could then begin again though much more slowly than Brunel had hoped. The sinking of the great cylinder had begun in June 1854, and it had reached the rock by the following February, but it was not until the end of 1856 that the pier was built above high water mark.

Today, Brunel's "great cylinder" would be called a compressed-air caisson. It has become the usual method of carrying out under-water engineering works of this kind, but in 1856 the successful completion of Saltash pier was a wonderful triumph of engineering skill and determination over every kind of difficulty and danger.

While this work was going on, the first of the two spans of the bridge was building on the bank of the Tamar, and on September 1, 1857, it was arranged to move it into position over the piers. The day was declared a general holiday. Church bells pealed; flags flew from every house in Saltash and from all the country round crowds flocked to the river banks to see the wonder performed. Specially built pontoons were floated under the span and took its weight as the tide rose. Then, at a quarter past one in the afternoon, Brunel climbed to a small signalling platform built on the very top of the great curving tube. A silence fell on the crowd as signal flags flickered from the platform and, in answer, cables tautened as winches on ships and shore swung into action.

While the people watched, awestruck, the great span, weighing more than a thousand tons, moved slowly and silently out into the Tamar and was swung until, with unerring skill, its two ends were brought directly above the piers. Then water was let into the pontoons so that they sank, leaving the span resting securely on the temporary caps of the piers.

As soon as it could be seen that the span was safely in place, the silence was broken by a storm of cheering and a band played "See the conquering hero comes" as Brunel came down from the high platform where, like the conductor of an orchestra, he had controlled the whole intricate operation.

With the span in position, it remained to raise it slowly to its full height by means of powerful hydraulic jacks, the piers being built up beneath it as it rose. Brunel's chief assistant, R. P. Brereton, superintended the floating and raising of the second great span in July 1858, and in the following spring the bridge was opened for traffic by the Prince Consort after whom it was named the Royal Albert Bridge. Its completion closed the last and greatest link in Brunel's long broad gauge main line from Paddington to Penzance.

Brunel himself did not attend the official opening of his great bridge; he was too ill. He crossed it once, lying upon a couch mounted upon a flat truck and gazing up at the huge iron tubes high overhead as one of Gooch's locomotives drew him slowly along. His railway career was ended and four months later he was dead, broken by the most ambitious of all his schemes – the *Great Eastern* steam ship. In black letters high above the pier arches of the bridge the directors of the Cornwall Railway placed to his memory the simple inscription which you may read there today: I. K. BRUNEL, ENGINEER, 1859. That is all. The bridge itself speaks for him.

8

The Race to New York

BY NOVEMBER 1854, WHEN HE WAS FORTY-EIGHT years old, Brunel had been responsible for building more than a thousand miles of new railways. By the end of his life the total must have been considerably more. You might think that this was more than enough work for one man to do in a lifetime and yet, while all this railway building was going on, Brunel was also responsible for three of the most remarkable steamships ever built. At the time of their launching, each in turn was the largest ship in the world.

At one of the earliest meetings of the Great Western Railway Company in London, one of the directors doubted whether the line to Bristol would ever be built because it was so long. To this Brunel replied: "Why not make it longer, and have a steamboat go from Bristol to New York and call it the *Great Western*?" At that time it was thought impossible to cross the Atlantic by steam because no steamship could carry enough coal for so long a voyage. So most of the railway directors laughed at what they thought was just a joke on Brunel's part. After all, even Brunel's father had once declared that 'steam cannot do for distant navigation'.

But one of the directors, a Bristol merchant named Thomas Guppy, knew Brunel better than the others and talked to him about the idea that evening after the meeting was over.

As Guppy suspected, Brunel's suggestion had been made perfectly seriously. He explained that whereas the load a ship's hull can carry increases as the *cube* of its dimensions, its resistance to motion – in other words the amount of power needed to drive it through the water – is a matter of area and therefore increases only as the *square* of its dimensions. So, if the size of a ship was doubled it would not, as was then supposed, need double the amount of fuel to drive it. On the contrary, said Brunel, a ship could be designed to carry enough coal for any given length of voyage and so transatlantic steam navigation was perfectly possible. It was simply a matter of building a ship of the right size, and as a result of this talk with Guppy the Great Western Steamship Company was formed in Bristol to do just this.

William Patterson, a Bristol shipbuilder, began building the *Great Western* in July 1836 to Brunel's design. Her immensely strong hull was built of oak sheathed in copper and she was to be propelled by paddle-wheels. When launched just a year later she was sailed round to the Thames with a tug escort to be fitted with engines of 750 h.p. which had been built by the firm of Maudslay, Sons & Field.

Rivalry between different ports and different shipping companies was so keen that as soon as the news of the *Great Western* became known two companies, one in

Liverpool and one in London, began to build steamships to compete with her. They felt that Brunel's ideas would very probably succeed and in that case they were determined that Brunel and the Port of Bristol should not have things all their own way. If they could snatch the credit of being first across the Atlantic they would do so, but it soon became clear to them that neither of these new ships could be completed in time. Not to be outdone, both companies chartered small cross-channel ships for the attempt, the choice of the London company being the little Irish channel paddle steamer *Sirius*.

By the time the *Great Western* entered the Thames to take on her engines, the *Sirius* was already there, being fitted out for her ocean voyage with coal bunkers of much larger size and other alterations. By a very small margin the *Sirius* won the desperate race to get the rival ships ready for sea. She put to sea on March 28, 1838. Three days later, early on the morning of March 31, the *Great Western* followed her with Brunel and his friends on board. They had to pick up passengers, coal and stores at Bristol before leaving for New York, but they were not unduly worried because they knew that the *Sirius* was due to make a similar stop at Cork, and Brunel was confident that his splendid new ship had the power and speed to overtake her rival. They had not travelled far, however, before a most alarming thing happened.

Brunel and his friends were on deck, the ship was off Leigh-on-Sea at the mouth of the Thames and going beautifully when suddenly smoke and flames began to pour from the boiler room. What had happened was that the heat of the furnace flue leading into the funnel had set fire to the felt lagging on the boilers and this in turn had set the underside of the deck alight.

Fearing the worst, the ship's Captain, Lieutenant James Hosken, R.N., ran the *Great Western* aground on

the Chapman sand while everyone ran to fight the fire. As was usual in such moments of danger and crisis, Brunel was well to the fore. He began to climb down the ladder into the smoke-filled boiler room, but the smoke was so thick that he did not see that the ladder had already partly burnt away. A charred rung collapsed under his weight and he fell heavily to the boiler room floor eighteen feet below. Injured and unconscious, he was rescued by his friend Captain Claxton who shouted for a rope and soon had him hauled by willing hands up to the deck. There he was wrapped in a sail, lowered into one of the ship's boats and rowed ashore to a farm on Canvey Island where he had to remain for several weeks. Meanwhile the fire had been put out and when the evening tide lifted her off the sand the *Great Western* was able to continue her voyage, but the mishap had lost her twelve vital hours.

Most unfortunately, too, a rumour reached Bristol that the ship had been completely destroyed by fire, with the result that all but seven of those who had booked to sail in her had cancelled their passages. Anxious though he was to beat the *Sirius*, Brunel was determined to take no risks. On his sick bed in Canvey Island he wrote a list of jobs that must be done before the ship sailed for America, and when these had been attended to there was a further delay of twenty-four hours caused by bad weather. So it was not until April 8 that the *Great Western* finally steamed away. It must have seemed to those on board that there was little hope left now of catching the *Sirius*.

In fact, although they did not know it, the *Sirius* had sailed from Cork, a whole day's steaming nearer New York than Bristol, on April 4. Shortly afterwards, when she was off the Blasket Islands at the westernmost tip of Ireland, the gallant little ship struck the same great storm which had delayed the departure of her rival. Conditions became so bad that both passengers and crew implored the Captain, Lieutenant Roberts, R.N., to abandon the attempt and put the ship about. But Lieutenant Roberts was a tough, skilful and determined seaman who insisted that the ship must go on, and go she did until, when three thousand miles had been covered, the coal began to run short and had to be eked out by burning some barrels of resin from the cargo.

However, to everyone's relief, the *Sirius* arrived off New York on April 22. She was given a great welcome when she docked next morning after nineteen days at sea, but it had been a very close call for she had only 15 tons of coal left in her bunkers.

The race, too, had been very narrowly won for, all unbeknown to the crew of the *Sirius* and the people of New York, in the early hours of that same morning the *Great Western* had dropped anchor off the mouth of New York river, fifteen days and five hours after leaving Bristol. Let an American reporter describe the scene as Brunel's ship steamed into harbour next day:

> The approach of the *Great Western* to the harbour, and in front of the Battery, was most magnificent. It was about four o'clock yesterday afternoon. The sky was clear – the crowds immense . . . Below, on the broad blue water, appeared this huge thing of life, with four masts and emitting volumes of smoke. She looked black and blackguard . . . rakish, cool, reckless, fierce, and forbidding in sombre colours to an extreme. As she neared the *Sirius*, she slackened her movements, and took a sweep round, forming a sort of half circle. At this moment, the whole Battery sent forth a tumultuous shout of delight, at the revelation of her magnificent proportions. After making another turn towards Staten Island, she made another sweep,

and shot towards East River with extraordinary speed. The vast multitude rent the air with their shouts again, waving handkerchiefs, hats, hurrahing!

The voyage of the *Sirius* had been a great and gallant feat of seamanship and in this respect her victory was well-deserved, but it had done nothing to prove the practicability of ocean steam navigation. Indeed the difficulty and danger of the feat tended to prove the opposite. Technically, the honours went to the *Great Western* and above all to the man who had designed her. In a voyage entirely without incident she had crossed the Atlantic with complete reliability and safety in a far faster time than her rival. Most important of all, when she docked she still had 200 tons of coal left in her bunkers.

There was no lack of passengers for the *Great Western's* return voyage to Bristol which was made in fifteen days. The two little channel steamers *Sirius* and *Royal William** soon returned to their normal duties when the rival ocean steamers *British Queen* and *Liverpool* entered service, but neither of them could equal the *Great Western* for speed and reliability. She made no less than sixty-seven Atlantic crossing in eight years and was the first ship to hold the Atlantic Blue Riband, her best crossings being thirteen days westbound and twelve days six hours eastbound.

* The Liverpool Company's challenger *Royal William* was an "also ran" in the Atlantic race. She arrived in New York on 24 July, 1838, after a voyage of eighteen days twenty-three hours from Liverpool. She was the smallest steam vessel ever to cross the Atlantic.

The *Great Western* had not completed her second voyage in 1838 before Brunel began to plan a bigger and better ship to be called the *Great Britain.* She would be nearly twice the size of the *Great Western* and propelled by engines of twice the power – 1,000 indicated horse-power. She was first planned as a wooden paddle steamer. To solve the problem of forging the huge paddle shaft James Nasmyth invented his steam hammer. But then Brunel changed his mind. He decided to make the hull of iron and to use a screw propeller instead of paddle wheels. There were small iron-hulled steamers afloat already, and a topsail schooner named the *Archimedes* had been fitted experimentally with an auxiliary engine and a screw propeller, but these two features had never before been combined in one ship or used upon so large a scale.

Like her predecessor, the *Great Britain* was built by William Patterson at Bristol, and the Prince Consort travelled down from Paddington in a special train driven by Gooch in order to name and launch her on July 19, 1843. The ship was so big that she stuck in the lock leading from the docks into the tidal river Avon and part of the wall had to be removed before she was able to pass through on her way to the Thames at Blackwall to complete her fitting out.

The *Great Britain* left on her maiden voyage to New York on August 26, 1845, and all went well, but on the return passage of her second voyage that year she broke her propeller and had to complete her journey under sail. During the following winter a new propeller was fitted and other improvements made before she began another series of monthly voyages. The 'teething troubles' due to the novelty of her design seemed to have been overcome and the Great Western Steamship Company felt every confidence in their fine ship when she left Liverpool for New York on September 22, 1846, with one hundred and eighty passengers on board, the largest number yet carried by any transatlantic steamer. Then came disaster.

A few hours after leaving Liverpool and in pitch darkness, the *Great Britain* struck. Passengers rushed from their cabins in terror as they heard the ship pounding over rocks with a frightful grinding noise. No wooden ship could have withstood such a battering; it would probably have broken up with great loss of life. But Brunel had given the iron hull such strength that everyone was able to remain on board in safety till dawn broke. Till then no one knew where they were. At the time the ship struck the Captain had supposed he was near the Isle of Man, but daylight revealed the mountains of Mourne. The *Great Britain* was high and dry on the beach in Dundrum Bay on the north-east coast of Ireland. The compass had been affected by the mass of iron in the hull and had given a false reading.

At low tide all the passengers and crew were brought to safety, but after one unsuccessful attempt to refloat her, she was abandoned by the Company as a total loss until Brunel himself managed to visit the scene in December. He was furious when he saw how little effort had been made to save his beautiful ship.

"The finest ship in the world," he told the Company, "has been left, and is lying, like a useless saucepan kicking about on the most exposed shore that you can imagine, with no more effort or skill applied to protect the property than the said saucepan would have received on the beach at Brighton . . ." Under his direction a timber breakwater was built to protect the ship from the winter storms, and in the following spring she was successfully floated off and towed to Liverpool for repair.

This disaster ruined the Great Western Steamship Company, but it was not the only reason why such a gallant venture failed. The saddest part of this story is the almost unbelievable stupidity of the Bristol Dock Company in neglecting the opportunities given to them by the genius of Brunel. The broad gauge main line to London and two of the finest ocean steamers in the world had given Bristol a great chance to become once again Britain's premier western port. But the Dock Company not only refused to make the improvement in the docks that ships of such size required but also made them pay terribly heavy dock charges. Because of this it became impossible for the ships to use their home port and they were driven away to Liverpool. This added greatly to the difficulties of the Great Western Steamship Company which was based at Bristol. It also meant that Liverpool and Southampton rose to become the two great ocean terminals of the steamship age.

The *Great Western* was sold to the Royal Mail Steam Packet Company and for ten years sailed regularly between Southampton and the West Indies. When the *Great Britain* had been repaired she was also sold and made no less than thirty-two voyages from Liverpool to

Australia and back in the next twenty-three years. Her last voyage was made in 1886 when she was damaged in a terrific gale off the Horn but managed to run for shelter to the Falkland Islands. There she was condemned and converted into a hulk for the storage of wool and coal. In 1937 the old ship was towed away to a quiet grave in nearby Sparrow Cove and there the remains of Brunel's almost indestructible hull may still be seen.

9

The Renkioi Hospital

THE CRIMEAN WAR GAVE BRUNEL AN OPPORtunity to use his extraordinary gifts in yet another different field. For most people mention of the Crimea calls to mind one name in particular – that of Florence Nightingale. The story of her journey, accompanied by a gallant little band of nurses, to the military hospital at Scutari and of the appalling conditions which she found there, is well known. Her reports on these conditions caused such a scandal at home in England that the Government fell and Lord Palmerston succeeded Lord Aberdeen as Prime Minister.

It so happened that at this time Brunel's brother-in-law, Sir Benjamin Hawes (he had married Brunel's elder sister Sophia) was Permanent Under Secretary at the War Office. In February 1855, he wrote to Brunel and asked him if he would be willing to design an improved hospital for the Crimea which could be built in England and then shipped out in parts for speedy erection on some chosen site. Brunel accepted immediately and went to work with such energy and speed that within six days he had not only designed the hospital but, on his own initiative, placed an initial contract for buildings to house a thousand beds. This provoked an outraged protest from the War Office Contracts Department. They were not accustomed to such indecent haste, and the placing of the contract without consulting them was highly irregular. But Brunel was not the man to be intimidated by bureaucrats, particularly when men's lives were at stake, and his reply was a stinging snub. "Such a course may be unusual in the execution of Government work," he wrote, "but it involves only an amount of responsibilty which men in my profession are accustomed to take." And he ended: "These buildings, *if wanted at all*, must be wanted before they can possibly arrive."

Although Brunel designed the hospital with such fantastic speed, the result showed all the care and forethought that he devoted to everything he undertook. Its general principle – a series of standard unit buildings linked by corridors – has governed the design of all portable or, as we now say, prefabricated structures of this kind from that day to this. Each standard unit housed forty-eight patients in two separate wards and was completely self-contained, having its own nurses' room, lavatory and wash-house complete with drainage system and fan ventilation. Similar wooden units housed the surgery and dispensary, but the kitchen, bakehouse and laundry units were of metal because of the fire risk.

It was at first proposed to erect this new hospital at Scutari, but a site near the village of Renkioi in the Dardanelles was eventually decided upon. In March Brunel warned Hawes that 1,800 tons of shipping space would very soon be required. A capable engineer would obviously be needed to prepare the site and superintend the erection of the hospital, and Brunel decided that the man for this job was John Brunton who was at that time building the railway from Dorchester to Weymouth.

The first Brunton knew of this was when he received a telegram one evening at his home at Dorchester ordering him to present himself at Brunel's office at 6 o'clock the following morning! The startled Brunton had no idea what this summons meant but he just had time to catch the night mail, and arrived in London in the darkness of a cold winter morning. What followed is best told in his own words: "At 6 o'clock I presented myself at 19 Duke Street Westminster." (Brunel used the house next door to No. 18 as offices.) "A footman in livery opened the door, and told me in reply to my enquiry that Mr. Brunel was in his office room expecting me. I was ushered into the room blazing with light, and saw Mr. Brunel sitting writing at his desk. He never raised his eyes from the paper at my entrance. I knew his peculiarities, so walked up to his desk and said shortly, 'Mr. Brunel, I received your telegram and here I am.' 'Ah,' was his reply, 'here's a letter to Mr. Hawes at the War Office in Pall Mall, be there with it at 10 o'c.' He resumed his writing and without a word further I left his office, and went and got my breakfast, more puzzled than ever as to the meaning and upshot of all this."

This little story of Brunton's helps us to understand how Brunel managed to get through such an enormous amount of work in his comparatively short life. It would be difficult today to find a professional man busy at his desk at six o'clock of a winter's morning, still less a servant on duty to admit visitors. Brunel never wasted an unnecessary word. In this case, as he well knew, Brunton would very soon learn "the upshot of all this" from Sir Benjamin Hawes whose responsibility it was to appoint Brunton. But when Brunton got the job the case was altered and he tells us later how, on revisiting Brunel, the details of the hospital scheme were explained to him most minutely and the difficulties in the way of his early departure abroad instantly solved.

For example, Brunton was worried about the future of his engineering pupil, Henry Waring, who still had eighteen months to serve with him. "Send him to my office," said Brunel. "I will take him without any premium." So, in an unbelievably short space of time, Brunton found himself on ship-board bound for the Dardanelles where he met Dr. Parkes, who had been appointed superintendent of the new hospital.

Letters from Brunel soon followed him. They are worth quoting because they reveal his powers of organization, the thought he devoted to every detail of his schemes and his working philosophy:

> All plans will be sent in duplicate . . . By steamer *Hawk* or *Gertrude* I shall send a derrick and most of the tools, and as each vessel sails you shall hear by post what is in her. You are most fortunate in having exactly the man in Dr. Parkes that I should have selected – an enthusiastic, clever, agreeable man,

devoted to the object, understanding the plans and works and quite disposed to attach as much importance to the perfection of the building and all those parts I deem most important as to mere doctoring.

The son of the contractor goes with the head foreman, ten carpenters, the foreman of the W.C. makers and two men who worked on the iron houses and can lay pipes. I am sending a small forge and two carpenters' benches, but you will need assistant carpenters and labourers, fifty or sixty in all . . . I shall have sent you excellent assistants – try and succeed. Do not *let anything induce you to alter the general system and arrangement that I have laid down.*

A few days later Brunton received another letter from Brunel which read:

Materials and men for the whole will leave next week. I will send you bills of lading for the five vessels: the schooner *Susan* and barque *Portwallis*, the sailers *Vassiter* and *Tedjorat* and the *Gertrude* and *Hawk* steamers. By the first named steamer, a fast one, the men will go . . .

I will only add to my instructions attention to closet floors by paving or other means so that water cannot lodge in it but it can be kept perfectly clean. If I have a monomania it is my belief in the efficacy of sweet air for invalids and the only point of my hospital I feel anxious about is this.

Five days after this the bills of lading for the

complete hospital arrived with a note which read:

> I trust these men will pull all together, but good management will always ensure this – and you must try, while you make each man more immediately responsible for his own work, to help each other – and to do this it is a good thing occasionally to put your hand to a tool yourself and blow the bellows or any other inferior work, not as a display but on some occasion when it is wanted and thus set an example. I have always found it answer.

Erection of the hospital began on May 21, a remarkable achievement when we remember that Brunel was not approached until February 16. But the work proved slower than had been hoped simply because, try as he would, Brunton could not find any reliable local labour, and the job had to be done almost entirely by the little band of eighteen men sent out from England. Nevertheless by July 12 the hospital was ready to admit three hundred patients and by December 4 its full total of a thousand beds were available.

The only tragic side to this story is that Brunel's help had not been called upon much sooner. All the misery and suffering of Scutari might then have been avoided. As it was, the Renkioi hospital with its spacious wards, its modern sanitation and air conditioning was only in full operation for a few months before the Crimean war ended and it had to be dismantled. Yet it was no labour in vain for in that short time fifteen hundred sick and sorely wounded men passed through its wards and of that number only fifty died, a very different record from that of Scutari where it was thought that a wounded man stood a better chance of recovery if he was left lying on the battlefield. Said Brunel in a final letter to Brunton:

> Everybody here expresses themselves highly satisfied with Everybody there and what we have done. I should wish to show that it was no *spirit* but just a sober exercise of common sense . . .

10

Building the Great Ship

WHILE RICH PEOPLE INVESTED THEIR MONEY IN new railways and new industries during the 1840s, the changes brought by rapid industrial revolution, by agricultural enclosures and an increasing population, combined to make life so difficult for the working man that the period has become known to historians as "the Hungry Forties". Many of them gave up the struggle by emigrating to the new world of America, but in 1851 Australia became the goal. It was the discovery of gold there and Australia's display of her agricultural wealth at the Great Exhibition in Hyde Park that brought this about.

There was a sudden boom in the Australian shipping trade and it soon became difficult to handle the increased traffic. Existing steamships such as the *Great Britain* had been designed for the Atlantic run and could not carry enough coal for the much longer voyage to Australia. Coal had to be specially sent out from South Wales to Cape Town so that the ships bound to or from Australia could refuel there and this was very expensive for the shipping companies. Just as he had earlier solved the problem of the Atlantic crossing, so now Brunel met this new difficulty by designing a ship which could carry enough coal for a voyage to Australia and back. This amounted to a voyage round the world. Once again it was simply a matter of proportions, but what proportions they were! The ship would be nearly 700 feet long and displace 32,000 tons – six times the size of the largest ship ever built at that time. Only Brunel would have dared to conceive such a gigantic ship, and only he had the persuasive power to convince others that it would be possible to build it.

In July 1852, the Great Eastern Steam Navigation Company accepted Brunel's design, and in the following December the contract for building the ship was placed with John Scott Russell whose shipyard was beside the Thames at Millwall. It was originally planned to name the ship *Leviathan*, but this was later changed to *Great Eastern.*

The hull of the *Great Eastern* "stands out as a milestone in the progress of building ships of iron and later steel". So wrote an expert on the history of shipbuilding. Like the Saltash bridge, its design was the sum of Brunel's experience, not only in developing his two previous ships but as a bridge-builder. Indeed there are affinities between iron bridges and iron ships, and Brunel undoubtedly profited from his experience in these two fields to the advantage of both. It was his bridge-building experience that made him look at a ship's hull as though it was a long girder, and insist upon giving it much greater strength lengthways than any earlier shipbuilder had though necessary.

But, as he realized, a ship's hull has to be a more

complicated structure than a bridge because of the changing stresses and strains it must withstand in a rough sea. For whereas a bridge girder is always supported at each end, a ship's hull may at one moment be supported in the same way between the peaks of two waves and in the next it may be supported only in the middle by a single wave. This means that the upper and lower members of the hull will be sometimes in compression and sometimes in tension.

To resist such stresses Brunel built immense strength into the hull of his last great ship and at the same time made it almost unsinkable. In the hull of the *Great Britain* he had introduced a form of double bottom to gain extra strength, but now he made this inner bottom watertight and extended it up the sides of the ship to a height of 5 feet above her deepest load line, the two layers of plating being 2·8 feet apart. In addition, a series of iron bulkheads built across the hull divided the ship into ten watertight compartments each 60 feet long.

It would have been impossible at that time to build a single steam engine large enough to propel a ship of such size, so Brunel decided to use two engines, one driving a 24-feet propeller and the other two huge paddle wheels 60 feet in diameter. The builder of the ship, John Scott Russell, was also responsible for the paddle engine, but the screw engine was the work of the famous firm of James Watt & Co., of the Soho Foundry, Birmingham.

How to build and launch into the Thames a ship of such size was a great problem, and after much discussion it was decided to build it along the river bank and then push it sideways down slipways to a point where the high tide would float it off. Extra land had to be rented at Millwall because the hull was much longer than the waterfront of Scott Russell's yard.

In all his railway works and in the building of his two previous ships Brunel had been fortunate in his associates, but in this his last great enterprise it was otherwise. So determined was Scott Russell to win fame for himself as builder of the great ship that he quoted much too low a price for doing so in order to be sure of getting the contract. As a result he was soon in financial difficulties, but for a while he concealed this from Brunel and kept himself in funds by claiming to have done, and receiving payment for, more work on the ship than he had in fact achieved. Also, to add insult to injury, he began telling Press reporters that he and not Brunel was the designer of the ship.

Finally, when Brunel became aware of what was going on, Scott Russell stopped all work on the ship and then went bankrupt, leaving no assets on which the unfortunate owners of the ship could make a claim. He offered to complete the ship if £15,000 a month was paid to him, but as more than two-thirds of the full contract price had already been paid to him, this impertinent suggestion was refused. Meanwhile the huge half-completed skeleton of Brunel's giant ship lay idle and rusting on the river bank. Sceptics declared that it was likely to remain there as a monument to human folly, because even if it was completed it could never be launched. For when it was ready for launching it would weigh 10,000 tons and in all human history no man had ever dared to attempt to move so great a weight.

Brunel once said bitterly that he wished a storm would break and sweep away the ship, and we may imagine how a man of his genius, energy and determination must have felt while solicitors, accountants and bankers argued about what should be done. Eventually, however, the matter was settled in the only possible way – the job was taken out of Scott Russell's hands altogether and Brunel himself would superintend the completion and launching of the *Great Eastern.*

So it was that in May 1856, Brunel faced the most difficult and hazardous task of his whole career as work began once more on his great ship. It would have been difficult enough in the best of circumstances, but it was made far more so by the fact that Scott Russell's men, who had previously worked on the ship, now refused to co-operate, doing all they could to make the job as slow and as costly as possible. But although in this indirect way Scott Russell was able to harass Brunel, he could not defeat the determination of so great a man. By the summer of 1857 the ship was ready for launching.

Unfortunately, however, owing to Russell's delaying tactics and to seemingly endless arguments over the property on which the ship stood, the building of the great cradles and launching ways by which Brunel planned to lower the ship into the river had not even begun at this time.

Finally the creditors of Scott Russell agreed to lease the property to the ship's owners, but for a rent so high that it could bankrupt them if the launch was long delayed. For this reason the work of preparing the launching ways had to be carried out with all possible speed, and Brunel was forced to attempt the launch with untried and insufficient equipment.

Two massive timber cradles had to be built to support the ship. Each was 120 feet long and was arranged to slide down a runway that sloped down for 240 feet into the river. It was essential that this runway should not yield beneath the weight of the ship, so hundreds of piles had to be driven into the mud and gravel of the river bank. On the heads of these were laid, first a layer of concrete two feet thick, then a triple layer of timber baulks, each a foot square, and finally the iron railway lines over which the cradles would slide. What was then a sloping river bank is now a vertical quay wall, but below the foot of this wall part of the old launching way can still be seen at low tide.

Construction of these cradles and ways was in itself a major engineering operation, and although the work was pushed forward with all possible speed it was not until the spring tides of early November 1857, that Brunel was able to attempt a launch, though not without misgiving. He had originally planned special hydraulic launching gear, but because of the financial difficulties in which Scott Russell had involved the company this had been judged too expensive, and Brunel had to rely on steam winches and capstans worked by teams of men to haul the ship down the ways. There was always the possibility that, once these winches had got the ship moving, she would begin to slide by her own weight. So to prevent her getting out of control she was secured by heavy chain cables to two enormous checking drums each 9 feet in diameter and 20 feet long.

It was in September of this same year that Brunel had superintended so successfully the floating out of the first span of his Saltash bridge, and he now decided to control the launch attempt in exactly the same way, using flags and numbers to signal his wishes to the different crews manning checking drums, winches and capstans from a platform built high on the deck of the ship. Success at Saltash had been due above all to the maintenance of perfect silence, order and discipline while the operation was carried out. True, large crowds had watched it, but they had been kept well out of the way.

Imagine Brunel's horror when he arrived at Millwall on November 3, the date fixed for the launch attempt, only to find that the whole yard was over-run by crowds of inquisitive people. He was furious, but there was nothing he could do for he discovered that, unknown to him, the company had been selling tickets admitting the public through the gates to see the launch. This being so he was forced to carry on although he would have much preferred to abandon the attempt.

Having climbed up to the high control platform, Brunel signalled all the winches and capstans into action, the wedges securing the cradles having first been knocked away. The effect was very dramatic, for the straining cables caused the huge iron hull to reverberate with a sound like a long roll of thunder. But the ship did not move until the power of two rams at bow and stern was brought to bear as well. Then the ground underfoot suddenly trembled as the bow cradle slid three feet. Because of the crowds and the general confusion it was not known whether it then stopped of its own accord or by the action of the crew manning the checking drum. The sudden movement had so alarmed a capstan gang on a barge moored in the river that they abandoned their bars in the belief that they were about to be overwhelmed. One jumped into a small boat and disappeared from the scene, rowing for dear life.

Meanwhile the men responsible for the winch controlling the heavy chain cable that was passed round the stern checking drum had abandoned their charge under pressure of the crowds, and an old Irishman named Donovan was actually sitting or standing on one of the winch handles. Suddenly the stern cradle slid four feet, the winch handles spun round, flinging the mutilated body of Donovan high into the air over the heads of the horrified crowd.

This accident brought things to a halt until the early afternoon by which time a heavy rain was falling from a leaden sky. After a brief effort that brought no further result, one of the steam winches stripped its gears, a capstan barge began to drag its anchor and Brunel decided to call it a day. This time no band played 'See the Conquering Hero" as Brunel came down from his platform. Instead a soaked and dispirited crowd elbowed their way out of the yard gates grumbling that they had been cheated out of their entrance money and laying the blame on Brunel.

Because the launching was dependent on the height of the tides, the next attempt was not made until

November 19, but no progress was achieved because the hauling cables parted. On November 28 Brunel tried again. Again the cables broke repeatedly and the fact that this time the ship was moved a further 14 feet down the ways was mainly due to the efforts of the hydraulic rams at the bow and stern. Brunel had looked to these rams merely to start the ship sliding and had hoped that, once started, the winches would keep her moving. But the repeated failure of the winches and their tackle now convinced him that he must rely on hydraulic power alone to launch his great ship. That it could do so he had no doubt, but it would be a very slow and painstaking business. For, unlike the special hydraulic launching gear he had originally hoped to produce, the rams he was now using were simply a form of jack with a very short length of thrust. Each time they had pushed the ship to the limit of their extension they had to be retracted, and then followed the slow and laborious task of inserting heavy packing pieces behind them so that they could make another push.

Throughout the month of November the ship was moved slowly forward, but the further she went the greater became the effort required to start her moving. Three times the rams burst their cylinders until, by mid-December, Brunel decided to stop the work until new, better and more powerful equipment had been obtained. He remembered the brothers Tangye, two clever mechanics whose work he had admired in Cornwall and who had lately moved to Birmingham. There, one dark winter's night, one of his assistants found the

Tangye brothers in a little workshop behind a baker's shop and gave them an order for the new hydraulic rams. The order proved the stepping-stone to fortune, transforming that small workshop into a famous engineering firm with the boast that "we launched the *Great Eastern* and she launched us."

By the New Year (1858) the new Tangye equipment had been delivered and with it had come the big hydraulic press which had been used by Robert Stephenson for lifting the tubes of his Britannia bridge. Brunel was confident that their combined thrust of 4,500 tons would be more than sufficient to launch the ship, but still ill-fortune seemed to dog the enterprise.

It was now the weather's turn to frustrate the hard-pressed engineer. By day thick fog hung over the river, stopping the work because it was impossible for the men working the rams to see Brunel's controlling signals. These foggy days were followed by bitter nights when fires had to be lit and tended through the long dark hours to prevent the frost from damaging the rams and the pumps that powered them. Yet in spite of these difficulties, by January 10 the ship had been pushed so far down the ways that she became partially waterborne at high tide, with the effect that her weight on the ways was reduced and progress was more rapid – so much so that Brunel suspended operations four days later in case the high spring tide due on the 19th of the month should float her prematurely.

He now determined to float the ship on the next high tides at the end of January, so once the tide of the 19th had passed she was pushed right to the end of the ways. Water ballast was then pumped into her double hull by a fire tender in order to hold the ship securely in the cradles until the great moment came. This was to be high tide in the early hours of the morning of January 30. On the day before, a Friday, young Henry Brunel was granted special leave from Harrow School so that he could be with his father to share an experience to be remembered all his life.

The weather that night was not promising, but four steam tugs were standing by to take charge of the ship. Brunel gave the order to the fire-float to begin pumping out the 2,700 tons of water ballast that held her down. Now, apart from reading the tide gauge every half hour, there was nothing that father and son could do but wait and hope. While the tip of his father's cigar glowed in the gloom of the little office by the slipway, Henry, unable to sleep, sat listening to the distant thudding of the pumps and to the boom of a rising wind that sent squalls of rain rattling against the windows.

Once again it seemed as if the fates were conspiring against the engineer for by the early hours of the morning it was blowing a full gale dead on the beam of the great ship. It would have been impossible for the tugs to control her in the teeth of such a wind so Brunel reluctantly ordered the pumps to be stopped. Later, the water ballast was pumped back, which was just as well because the tide rose exceptionally high and the ballast was only just sufficient to hold the ship.

All that day and well into the night the storm raged,

but a little after midnight the rain stopped at last. The wind moderated and changed from south-west to north-east. If the ship could not be launched on the Sunday morning tide (January 31) she would have to wait until the next springs, so Brunel determined to make the attempt. At 3.30 a.m., exactly twenty-four hours after they had been stopped, he ordered the pumps to start pumping out once more. By 6.0 a.m., when he woke Henry from an uneasy sleep, the sky was bright with stars. At last the sun rose and the men, roused by messengers, arrived to stand by the rams and steam winches for the last time. As the tide rose that morning the ship was pushed cautiously forward until the cradles were right off the end of the ways.

At one o'clock Brunel's wife, Mary, and his sister, Sophia, arrived at the yard in a carriage to see the launch. They were just in time for just twenty minutes later the stern, which had been deliberately pushed further than the bow, was seen to be afloat. Brunel immediately signalled the for'ard steam winch to haul out. At 1.42 p.m. precisely the great bows lifted gently on the tide. The *Great Eastern* was afloat at last.

Brunel and his family at once went on board and the four tugs then began to manœuvre the huge hull clear of the shore and out into the river. As they did so the two timber cradles that had supported the ship for so many months shot to the surface from beneath her with a swirl and a roar of waters. By seven o'clock the ship had been safely moored at Deptford and Brunel stepped ashore to the cheers of his men. He had achieved what most people had thought impossible in the face of every kind of difficulty that man and nature could put in his way, but his victory had been won at terrible cost.

11

The Last Months

BRUNEL HAD FLUNG HIMSELF INTO THE GREAT Eastern project with all the unsparing energy that he had once given to his father's Thames Tunnel. He had spent long days, and sometimes nights as well, at the ship-yard, often under appalling weather conditions. He had been little more than a youth in the Thames Tunnel days, but now he was a man in late middle age. Also, to add to the bodily strain and hardship, there was all the worry that John Scott Russell had caused him. The tremendous impetus of his will-power had driven him forward until he had launched his ship, but now that this was done that force was spent. He realized that he had driven himself too hard and was seriously ill. His doctors diagnosed Bright's disease, or nephritis as we now call it, and insisted that he must go abroad for a complete change and rest.

Ill though he was, Brunel was very reluctant to leave England. Indeed this reluctance was probably due to a knowledge that he had not very long to live, combined with a passionate determination to finish his last great task – to complete his ship and put her to sea. True that she was now afloat, but the £732,000 which had been spent on her had ruined the Company. It was estimated that a further £172,000 would have to be spent in order to fit her out ready for sea, and until that money could be found the *Great Eastern* would remain a useless hulk. While Brunel and his wife were away at Vichy and in Switzerland the Company tried to raise more money but with no success, so that when he returned to England in September 1858, he found his great ship still lying forlornly at Deptford just as he had left her four months before.

In November a new concern called the Great Ship Company was formed to buy the ship and complete the work on her, and the old Eastern Steam Navigation Company was afterwards wound up. Brunel was determined to protect this new Company from the kind of troubles that had ruined the old. He prepared most detailed and careful specifications and estimates for all the work needed to complete the ship, and he begged the Company to be sure that whoever undertook this work signed a contract binding him to follow these specifications in every detail.

These specifications had just been put out for tender when Brunel again had to go abroad, taking his son Henry with him as well as his wife, for his doctors had told him that he must spend the winter in a warm climate. This time Egypt was their destination, but they had got no further than Lyons on December 6 when the news reached Brunel that two tenders had been submitted for the work on the ship. One of these came from a firm named Wigram & Lucas and was based on Brunel's specification. The other was from John Scott

Russell who refused to accept the specification. Brunel at once wrote to the Company imploring them not to do business with Scott Russell, but in vain – his tender was accepted upon his own terms.

How Scott Russell, so soon after his bankruptcy, was able to tender and why the Company was prepared to employ him again on the ship are mysteries which may never be properly explained, but we must remember that, with Brunel away, Scott Russell could claim that he knew more about the ship than any other man in England. But the knowledge that the man who had caused him so much trouble before was back again was a terrible blow to Brunel. He and his family made an adventurous voyage up the Nile in a converted native boat, and on Christmas Day he dined in Cairo with Robert Stephenson, who had also ruined his health through overwork. But all the while his thoughts kept turning to his great ship, wondering what was happening on board.

When Brunel finally returned to England and stepped on board his ship again in May 1859, it was to find his worst fears confirmed. History had repeated itself. A bitter dispute was going on between Scott Russell and the Company over payment and work done, and the effect of this was that work on the ship was in chaos and far behind time. Perhaps because the Egyptian sun had tanned his face, Brunel's friends told him how much healthier he was looking, but he knew otherwise. He knew that death could not be far off, but this did not alter his determination once again to take command himself in a last desperate effort to make up for lost time and finish the job. On the previous occasion it had been a race against financial ruin; now it was a much more gallant race against approaching death, though it is doubtful whether anyone else realized this.

Nothing could rob Brunel of his drive and magnetism. The effect of his presence was electrifying and the chaos soon became order. Then, at the end of July, Brunel became so ill that he was forced to take to his bed at his home in Duke Street. Taking advantage of his absence, Scott Russell made a speech at an important dinner on board the ship following which the newspapers reported that: "the merit of the construction of the ship and her successful completion is owing entirely to the untiring energy and skill of Mr. Scott Russell."

But if Scott Russell supposed he had seen the last of Brunel he was mistaken. By mid-August the unconquerable engineer was back again, having rented a house in Sydenham so as to be nearer the ship. He had become a shrunken, pitiful figure now, moving slowly and painfully about the decks with the help of a stick, but still he came on board daily and still there could be no doubt as to who was in command as his cool, incisive orders poured out. Such was their effect that before the end of the month a sailing date had been arranged.

On Wednesday, September 7 the *Great Eastern* would steam down the Thames estuary to the Nore where she would adjust her compasses before going on to Weymouth. Brunel had reserved his cabin for this first short voyage which would make the triumphant completion of all his self-destructive efforts over the past five years. He had conquered every difficulty; nothing, he thought, could rob him of that triumph now, but on September 5 he collapsed on the deck. A sudden stroke had left him paralysed but still conscious. As he was carried ashore and home to Duke Street he realized that his great ship would sail without him after all.

Brunel fought even against this mortal blow, for a

few days later he was dictating letters in his bedroom to his faithful clerk Bennett. The last letter he ever signed was addressed to the Superintendent of the Great Western Railway Works at Swindon. It was a request that the men of the works, who had always been so loyal to him, should be given special railway passes and time off so that they could see his ship while she lay off Weymouth.

While Brunel was dictating this letter the *Great Eastern* was steaming majestically down the English Channel at 13 knots after moving down to Purfleet on

September 7 and to the Nore on the following day. At five minutes past six o'clock she was off the Dungeness lighthouse. A little group of passengers was standing in the bows admiring the way the huge ship was forging her way so smoothly through seas so rough that they had sent small craft scurrying to shelter.

Suddenly an unbelievable thing happened. There was a tremendous explosion. The group at the bow saw the first of the ship's five tall funnels suddenly launch itself into the air upon a cloud of steam like some enormous rocket. This is how one of them afterwards described the scene: "The forward part of the deck appeared to spring like a mine, blowing the funnel up into the air. There was a confused roar amid which came the awful crash of timber and iron mingled together in frightful uproar and then all was hidden in a rush of steam. Blinded and almost stunned by the overwhelming concussion, those on the bridge stood motionless in the white vapour till they were reminded of the necessity of seeking shelter by the shower of wreck – glass, gilt work, saloon ornaments and pieces of wood which began to fall like rain in all directions."

The two leading funnels passed through the Grand Saloon on their way up from the paddle engine boiler room, their presence concealed by large gilt mirrors. The explosion completely wrecked this saloon. Mercifully, no one was in it at the time, but in the boiler room six unfortunate stokers were fatally scalded by the escaping steam. It says a great deal for Brunel's design that in spite of this major disaster the great ship scarcely faltered on her course to Weymouth. The paddle engines continued to work on two of their four boilers with the addition of steam supplied from the screw engine boiler room.

Meanwhile Brunel lay dying at Duke Street, clinging stubbornly to life until news of his ship could reach him. He had felt so confident that this first voyage would prove a triumph to repay him for every misfortune, but instead a messenger arrived from Weymouth with the news of this crowning disaster. It was as though, in planning this huge ship, so far ahead of its time, Brunel had tempted fate too far and that fate was determined to be revenged on him. No one could have fought against fate more courageously than he had, but this final blow was too heavy for him and on the evening of September 15 he died. He had lived fifty-three years, a shorter life than most men hope for, but what crowded, exciting and fruitful years they had been!

The same newspapers that carried the news of Brunel's death also contained the report of the inquest at Weymouth on the unfortunate stokers who had lost their lives in the explosion. What had caused it? Where the funnels passed between the decks of the *Great Eastern* they were surrounded by water-filled jackets or casings. These prevented the funnels from giving out too much heat into the saloons and they also acted as feed-water heaters, in other words the water in them was used to supply the boilers. To supply a boiler with hot instead of cold water saves fuel and for this reason engineers sometimes call feed-water heaters economizers.

Years before, the fire on the *Great Western* had shown

Brunel how hot the lower part of a ship's funnel could get at that time and this was why he adopted these water casings, first on the single funnel of his *Great Britain* and finally on the five funnels of the *Great Eastern.* The casing ended at the level of the top deck, but a long pipe extended upwards from them to the full height of the funnels. These pipes were intended to be open so that there would be no risk of steam pressure building up in the casings, but so that the casings could be given a water pressure test to make sure that they were sound, stop taps had been fitted to the bottom of each pipe.

These taps were only intended to be a temporary arrangement, and when James Watt & Co., who were responsible for the screw engines and boilers, had tested their three water casings they removed the taps and so left the pipes open. John Scott Russell was responsible for the paddle engines and boilers and his men did not remove their two taps. This would not have mattered so long as they were open, but for some reason that nobody was able to explain, when the ship left the Nore, both taps were closed. As a result, steam pressure steadily built up in the water casings until the foremost one burst.

It became clear at the inquest that one of Russell's men, Arnott by name, realized the cause of the disaster as soon as the explosion happened. For he immediately sent an assistant up on deck with a key to open the second tap. The steam pressure in the second casing was harmlessly released up the pipe and the danger of a second explosion, which might have been even more disastrous than the first, was averted. In this you may say that Arnott showed great presence of mind, but his evidence at the inquest was very unsatisfactory and the fact that he knew the cause of the explosion the moment it happened may be the key to the mystery of the closed taps.

From the earliest days of steam power at sea down to the present day it has been the practice of the engineers who have built a new ship to take her out on sea trials and not to hand over responsibility to her owners until those trials have been completed successfully. So, when the *Great Eastern* put to sea, engineers from James Watt & Co. were in charge of the screw engines and Scott Russell's men were working the paddle engines. Russell himself was on the bridge giving them orders, yet at the inquest he disclaimed all responsibility and said that he had come on the voyage simply out of personal interest.

After this, no one connected with the ship could be in any doubt about the kind of man Scott Russell was. If, as seems likely, he had hoped to succeed Brunel as engineer to the Great Ship Company he was disappointed. After a final row, Russell's association with the *Great Eastern* ended for good. Brunel's great railway friend, Daniel Gooch, became engineer to the Company.

As a passenger liner the *Great Eastern* was never a commercial success. The Australian trade for which she was designed had declined by the time she went into service. Another blow to her was the opening of the Suez Canal. This shortened the sea route to Australia and the Far East, but as the canal was first built, the

Great Eastern, with her immense breadth over paddle wheels of 118 feet, could not pass through it. So the great ship was put on the Atlantic run like her two predecessors and for this service she was, at that date, far too big. Her greatest and most profitable work was done as a cable layer and for this Gooch was responsible. He chartered the ship to the Telegraph Construction Company and sailed with her on both the attempts to lay a telegraph cable across the Atlantic.

When the *Great Eastern* sailed with the first cable from Valencia in Western Ireland in July 1865, Gooch sat in his cabin and wrote: "The work has the best wishes and prayers of all who know of it. Its success will open out a useful future for our noble ship, lift her out of the depression under which she has laboured from her birth, and satisfy me that I have done wisely in never losing confidence in her; and the world may still feel thankful to my old friend Brunel that he designed and carried out the construction of so noble a work."

This first attempt failed when, to Gooch's bitter disappointment, the cable broke in deep water in mid-Atlantic and could not be recovered. Next summer they tried again and Gooch thought it a good omen when he saw that the Cork tug which came to assist their departure from Ireland was named *Brunel*. It *was* a good omen. Not only was the cable successfully landed at Heart's Content Bay in Newfoundland, but having done this the *Great Eastern* steamed out into the Atlantic again where the cable that had been lost twelve months before was recovered and joined.

It was Gooch who sent the first message through the

cable from the new world to the old. It read: "Our shore end has just been laid and a most perfect cable, under God's blessing, has completed telegraphic communication between England and the Continent of America."

For this achievement Queen Victoria made Gooch a baronet when he returned to England but, as he would have been the first to acknowledge, the real credit was due to the vision of the man who had died seven years before. For no other ship in the world could have laid that cable, as Brunel well knew. When the originator of the Atlantic cable scheme, the American Cyrus Field, first came to England, Brunel had taken him to see the huge unfinished hull at Millwall and said: "There is your ship." How right he was, for after this first triumph his *Great Eastern* went on to weave a web of cables round the world. So the great ship that had been ridiculed as a white elephant, that had seemed dogged by misfortune and had brought death to her creator, justified both herself and her builder in the end. And for the man who had built the first successful transatlantic steamship, the *Great Eastern* could not have done so in a more fitting way.

Some people think Brunel's career was a failure; that his schemes were extravagant and over-bold. Certainly he was not successful in the worldly sense, because although he made a lot of money as a railway engineer he lost most of it on the *Great Eastern*. His broad gauge, the atmospheric railway, the Great Western Steamship Company and the *Great Eastern* were all commercial failures. Even his father's Thames Tunnel, on which he spent his youthful energy, never became a vehicle tunnel as the Brunels intended but stood almost useless, taking foot passengers only, until it became a railway tunnel in 1865. Even his first work, the Clifton suspension bridge, was never finished in his lifetime.

Had Brunel, like other engineers of his day, been more cautious, had not thought upon so grand a scale, had not set his sights so high or been so eager to try out new ideas, he might have been more immediately successful. He would have saved himself a lot of anxiety, would have avoided much bitter criticism and might have lived to old age, rich and loaded with honours.

But would he really have been so great a man? Caution is not a mark of greatness but of timidity. Truly great men are those who, without fear, exercise to the utmost the power within them even though, by so doing, they may destroy themselves. They are the true leaders, blazing a new trail for their successors. That Brunel was of that company there can be no question. The broad gauge *Flying Dutchman* thundering down his splendid road to the west, and his three magnificent ships led the world into a new age of speed.

Only the ignorant and the jealous, the penny-wise and the faint-hearted, disliked and derided Brunel. Those who were nearest and dearest to him knew him for the genius that he was. No man could claim to know him better than Daniel Gooch. They had fought the battle of the gauges together; together they had made the Western Railway great and when Brunel died Gooch wrote in his diary: "On the 15th September I lost my oldest and best friend . . . By his death the greatest of England's engineers was lost, the man with the greatest originality of thought and power of execution, bold in his plans but right. The commercial world thought him extravagant; but although he was so, great things are not done by those who sit down and count the cost of every thought and act."

INDEX

THE LANDSCAPE TRILOGY

L.T.C. ROLT

For the first time, Rolt's three autobiographies are published together in one paperback volume. They tell of his rural childhood, his engineering apprenticeship, his passion for both motor racing and engineering and the garage he ran which specialised in veteran and vintage cars. The book continues with the fulfilment of Rolt's dream to convert the narrow boat *Cressy* into a floating home for his voyages through the secret green water-lanes of England and Wales during the Second World War and the blossoming of his writing career. Imbued with the author's love of England and his intense feeling for the beauty of the English countryside, *The Landscape Trilogy* reveals a landscape populated not only by men but also by the machines with which Rolt was fascinated.

L.T.C. Rolt was born in 1910 and became a pioneer of canal and railway preservation. He was the first to give literary shape to the subjects of the Industrial Revolution, and his biographies of Brunel, Stephenson, Telford and Newcomen are now regarded as classics. He died in 1974.

25 B/W ILLUSTRATIONS

ISBN 0 7509 4139 1

THE BALLOONISTS: THE HISTORY OF THE FIRST AERONAUTS

L.T.C. ROLT

Foreword by DON CAMERON

This book is a complete history of balloons and the intrepid men who flew in them. Beginning in 1783, the year in which balloons first took flight, it ends in 1903, the year in which the Wright brothers' first heavier-than-air flight at Kittyhawk changed the history of aviation for ever. The exploits of balloonists attracted the attention and admiration of the masses like nothing before: within weeks of the first flights, its form featured in designs of wallpaper and fabrics, in jewels and on snuff boxes, and as balloon clocks and chandeliers. The aeronauts themselves became heroes of their time. Covering the first flight by the Montgolfier brothers, in a balloon of paper and cloth, to the first Channel crossing by air, showmen aeronauts, female aeronauts, efforts to cross the Atlantic and the use of balloons in war, this is a wholly fascinating and riveting book. It includes lively extracts from journals and contemporary accounts, as well as engravings of the period.

56 B/W ILLUSTRATIONS

ISBN 0 7509 4202 9

NARROW BOAT

L.T.C. ROLT

In 1939, L.T.C. Rolt and his first wife Angela set out on a 400-mile journey through the canal network of the English Midlands. First published in 1944, this beautifully written book is the story of the canal people they met and the sights and sounds of the English countryside along the way.

55 B/W ILLUSTRATIONS

ISBN 0 7509 0806 8

A CANAL PEOPLE: THE PHOTOGRAPHS OF ROBERT LONGDEN

SONIA ROLT

Robert Longden's photographs of the narrow boat community at Hawkesbury Stop are an evocative reminder of a lost way of life. The book provides a rare insight into the community who worked the waterways when it was still a way of life for many, and will appeal not only to canal enthusiasts, but to anyone interested in Britain's social and industrial heritage.

130 B/W ILLUSTRATIONS

ISBN 0 7509 1776 8